SpringerBriefs in Applied Sciences and Technology

Safety Management

Series Editors

Eric Marsden, FonCSI, Toulouse, France

Caroline Kamaté, FonCSI, Toulouse, France

Corinne Bieder, FonCSI, Toulouse, France

The *SpringerBriefs in Safety Management* present cutting-edge research results on the management of technological risks and decision-making in high-stakes settings.

Decision-making in high-hazard environments is often affected by uncertainty and ambiguity; it is characterized by trade-offs between multiple, competing objectives. Managers and regulators need conceptual tools to help them develop risk management strategies, establish appropriate compromises and justify their decisions in such ambiguous settings. This series weaves together insights from multiple scientific disciplines that shed light on these problems, including organization studies, psychology, sociology, economics, law and engineering. It explores novel topics related to safety management, anticipating operational challenges in high-hazard industries and the societal concerns associated with these activities.

These publications are by and for academics and practitioners (industry, regulators) in safety management and risk research. Relevant industry sectors include nuclear, offshore oil and gas, chemicals processing, aviation, railways, construction and healthcare. Some emphasis is placed on explaining concepts to a non-specialized audience, and the shorter format ensures a concentrated approach to the topics treated.

The *SpringerBriefs in Safety Management* series is coordinated by the Foundation for an Industrial Safety Culture (FonCSI), a public-interest research foundation based in Toulouse, France. The FonCSI funds research on industrial safety and the management of technological risks, identifies and highlights new ideas and innovative practices, and disseminates research results to all interested parties. The series does not accept unsolicited submissions. For more information: https://www.foncsi.org/

FONCSI

Foundation for an
Industrial Safety Culture

Corinne Bieder · Caroline Kamaté · Hervé Laroche ·
Eric Marsden · Jean Pariès · René Amalberti
Editors

Living with the New Safety Landscape I

Rethinking Safety Management in the Context of Cascading Uncertainties

Springer

Editors
Corinne Bieder
FonCSI
Toulouse, France

Caroline Kamaté
FonCSI
Toulouse, France

Hervé Laroche
FonCSI
Toulouse, France

Eric Marsden
FonCSI
Toulouse, France

Jean Pariès
FonCSI
Toulouse, France

René Amalberti
FonCSI
Toulouse, France

ISSN 2191-530X ISSN 2191-5318 (electronic)
SpringerBriefs in Applied Sciences and Technology
ISSN 2520-8004 ISSN 2520-8012 (electronic)
SpringerBriefs in Safety Management
ISBN 978-3-032-29335-0 ISBN 978-3-032-29336-7 (eBook)
https://doi.org/10.1007/978-3-032-29336-7

This work was supported by Fondation pour une Culture de Sécurité Industrielle.

This Springer imprint is published by the registered company Springer Nature Switzerland AG
The registered company address is: Gewerbestrasse 11, 6330 Cham, Switzerland

Preface

Safety has made significant progress since the advent of high-risk industries. Major accidents are fortunately rare nowadays and there are a number of clues indicating that it is more and more taken for granted. Yet, if the historical approach to safety has made these safety results possible, the recent evolution and transformations of the world, and those to come, are likely to challenge current safety models, approaches, and practices, beyond the need for minor adjustments at the time. Indeed, major changes are ongoing at an unprecedented pace. Some are massive; they change the whole world or parts of it in many respects including safety in the way it is perceived but also the way it can be maintained or enhanced. Some changes are more local and affect safety in a more direct and obvious way. This acceleration of the transformations affecting safety urges the safety community to think ahead and get prepared to face unprecedented situations and expectations in the future, including a relatively near one.

This book in two volumes embraces the topic of the future of safety management and governance from as wide a perspective as possible to shed light on the current situation, its destabilization in the face of the many changes underway and to come and the possible ways forward to live with the inevitable uncertainties, complexities, inconsistencies. Drawing up a global picture is the purpose of the first chapter of the first volume meant to be the backbone of the two volumes and authored by Foncsi experts. The other chapters develop more detailed analyses and proposals on selected aspects addressed by the backbone chapter to substantiate the global view. These chapters are authored by international experts of these topics based on their research work and experience. Volume 1 starts with the backbone chapter *The crumbling facades of industrial safety*, before addressing new frames for a changing world through seven complementary chapters. Volume 2 addresses the organizations and safety challenges in a first part comprising seven chapters, then provides illustrations of destabilizations, and ways forward through cases and examples in a second part comprising four chapters, before standing back again and developing a wide reflection on the safety governance regimes in its final chapter *Regimes of Fractured Objectivity: Beyond the Envelope of Technology Safety Governance.*

The very recent changes especially in the political or geopolitical domain (e.g., dismantling of regulatory bodies or research labs) tend to make any reflection on the topic almost immediately outdated, at least partly. The content of this book is no exception. It already evolved from what it was at the beginning of this initiative and would probably not be exactly the same should the book be issued a few months later. Nevertheless, the aim is not to be completely accurate (who could pretend to be so when it comes to addressing the future), but rather to start a broad and collective reflection—being aware of the uncertainties and remaining modest—and trigger initiatives to keep up with an evolving world.

Toulouse, France

Corinne Bieder
Caroline Kamaté
Hervé Laroche
Eric Marsden
Jean Pariès
René Amalberti

Contents

Part II New Frames for a Changing World

Part I
Introduction

Chapter 1
The Crumbling Facades of Industrial Safety

Corinne Bieder, Hervé Laroche, and Eric Marsden

Abstract The world is undergoing major transformations at unprecedented pace and scale. Climate change, AI, but also radical political decisions regarding industries and research or the volume and circulation of information are now part of today's landscape which keeps evolving. Even if some of these changes might seem far away from industrial safety, all of them affect it either directly or through its ecosystem. In fact, they challenge safety management and governance in an unprecedented way. Far beyond the toolbox currently in use in safety practices, they question the very social and political foundations of safety. After analyzing the current status quo on safety management and governance through the lens of the "organized hypocrisy," this chapter explores the many ways in which it is and will be destabilized by the multifaceted evolution underway. Three strategies with different levels of ambition but also requiring action at different levels are proposed to live with this new safety landscape.

Keywords Safety management · Safety governance · Organized hypocrisy · Organizational façades · Safety as demonstrated · Safety as practiced

1.1 Introduction: Times of Crisis for Safety?

The world is undergoing huge transformations that challenge the way safety has been managed over the past decades. Technology has an increasing impact on our work and our private life. Climate change is not only having direct impacts on lives, the environment, infrastructures, or facilities with unprecedented extreme events,

C. Bieder (✉) · H. Laroche · E. Marsden
Foncsi, Toulouse, France
e-mail: corinne.bieder@foncsi.org

H. Laroche
e-mail: herve.laroche@foncsi.org

E. Marsden
e-mail: eric.marsden@foncsi.org

C. Bieder et al. (eds.), *Living with the New Safety Landscape I*,
SpringerBriefs in Safety Management, https://doi.org/10.1007/978-3-032-29336-7_1

but also indirectly through the upsetting of geopolitical balances and challenge of democratic regimes that are struggling to keep up with the rapid changes. If some evolutions were anticipated, including in the area of industrial safety, others and even more so their combined effects were not, at least in their magnitude and pace.

Some aspects that became blatant recently were not even thought of. It is the case of the immediate effects of radical political decisions, for example the closure and dismantling of research labs and federal agencies in the USA. These experiences add to the complexities and uncertainties that already made anticipation an illusion. Further, recent political discourses have reinforced the confusion between science, opinions, intuitions that was already observed at lower power level in society, with amplified effects. Most of these phenomena were unthoughts until recently.

The changes underway have a huge destabilization potential, including for the safety of high-risk industries or reliability of critical infrastructures, and some impacts can already be observed.

The ongoing transformations in the world will alter the contexts of trade-offs between safety and other dimensions of performance, such as productivity, social aspects, or sustainability. This will inevitably affect the outcomes. This contextual sensitivity is illustrated indirectly by Tony Licu in his chapter *Living Safely in a Changing Operational Environment* (Volume 2). He recounts how his perspective on the same safety issues evolved as he transitioned from a safety manager to an operational manager role within a large Air Traffic Management organization.

This chapter starts with an analysis of the current safety management and governance approaches and practices through the lens of the organized hypocrisy framework. We contend that, due to their magnitude and novelty, analyzing the current challenges for safety requires a societal approach. It then explores the turbulences ahead through all the ways in which the current status quo on safety management and governance is being and will even more in the future be destabilized. Based on these elements, a last section suggests ways forward for safety management and governance, and possibly new forms of hypocrisy, to live with the transformations underway and to come.

1.2 Safety Management and Governance: The Hypocrisy We Are Living In

1.2.1 Comfortable Discrepancies: Safety as Demonstrated Versus Safety as Practiced

The current state of affairs in safety management and governance is characterized, at the same time, as relatively stable and smooth, and as laden with internal divergence. In democratic countries where high-risk industries expanded dramatically during the twentieth century, discrepancies between the way safety is defined and ensured through regulatory requirements, processes, and procedures and the way safety is actually achieved in practice have long been identified and discussed. The former builds on an abstract view of operational situations and aims to provide, a priori, a safe response to each of them "on paper," based on models of anticipation and control. This "good old safety model" sees safety as relying on "reliable equipment, good procedures and processes, [and] well-behaved operators" (Bourrier and Bieder 2017, p. 3) and, since the 1990s, also on "well-designed organizations" (ibid.) and effective safety management systems. Within this framework, compliance with rules, processes, and prescriptions is central to safety management, as well as to the attribution of responsibility in the event of an accident.

Although safe design of equipment and infrastructure and compliance with well-designed regulations and procedures have played a major role in improving safety over the twentieth century, social scientists have shown for decades that compliance is not sufficient: Events and situations that were not or not accurately anticipated by system designers require appropriate responses from front-line workers who use their accumulated skills and experience to identify and characterize the singular situation they are facing and implement an appropriate response. These initiative behaviors are also needed to interpret and adapt the procedural framework to the variety of situations encountered day-to-day (including in some circumstances by deviating from the official procedure). The same applies at the level of safety management systems where the requirements, although commonly defined at a higher level than that of procedures, necessitate interpretation and adaption to the specific characteristics of a high-risk organization.

The "good old safety model" does not only dictate the way safety is defined and designed. It also underpins the way high-risk organizations demonstrate, both internally and to safety authorities, that their operations are safe. It is the official discourse, socially spread, about the way safety is controlled. This has important implications that have been largely overlooked. Exploring the consequences of the alignment between the "good old safety model" and the model inspiring the social governance of safety is the main purpose of this chapter. Rather than discussing the discrepancies between the way safety is defined and the way safety is practiced, we will shift the angle to examine the discrepancies between safety as demonstrated and safety as practiced.

In this chapter, we use the term "safety as demonstrated framework" to refer to the set of activities related to the anticipation of possible threats, the design of safety barriers intended to maintain control even when threats to safety arise, the development and improvement of operating procedures that cover all anticipated operational conditions, the development and improvement of organizational processes dedicated to managing safety. This approach involves the certification and "safety case" activities when companies design a facility or infrastructure, as well as all the activities throughout the organization's operational "life" whose purpose is to demonstrate to the safety authority that (future) operations do not and will not pose risks that society cannot tolerate.[1] Improving the design of technical systems or operational procedures based on operational experience, or of organizational processes to manage safety, is also part of the "as demonstrated" framework. All these "safety as demonstrated" activities rely on anticipation and involve formalization of a priori safe responses—individual or organizational—based on the knowledge available at the time.

We use the term "safety as practiced" to refer to the actual practices, both individual and organizational, used to produce safety, combining compliance with the "safety as demonstrated" framework where applicable in the context of the day, with adaptive capacity and initiative behaviors, which despite their less formalized nature are also essential to safe operation of complex systems operating in real-world environments in which full anticipation is impossible. Concepts and theories have been developed to describe the adaptive mechanisms at both individual and organizational levels, including High Reliability Organizations (HROs), Safety-I and Safety-II, and resilience (La Porte 1996; Hollnagel et al. 2006; Hollnagel 2018). Similar developments can be observed at the level of safety governance. Companies follow regulatory requirements, develop safety cases that demonstrate that all anticipated hazards are controlled, put in place safety management systems, and implement inspection and audit mechanisms to ensure that the procedures are respected. Safety-mature organizations also go beyond regulatory requirements and their recognized limitations to manage and enhance safety through practices that are not officially acknowledged by authorities as contributors to safety.

"Safety as demonstrated" and "safety as practiced" partly overlap in high-risk industries and critical infrastructures, as illustrated in Fig. 1.1. Indeed, to fulfill legal and regulatory safety requirements, high-risk organizations develop and maintain/update/upgrade processes and procedures that not only support the safety demonstration to the authorities but also internally guide decisions and actions at individual and organizational levels. These references and prescriptions result from the combination of safety concerns with other organizational stakes. The overlap between "safety as demonstrated" and "safety as practiced" varies with industry sectors, organizations and authorities' philosophies, facilities, time, etc.

[1] They form a rational-legal framework in the sense of Weber, in which authority regarding safety matters flows from stable rules implemented by a bureaucracy that is subject to judicial and democratic oversight.

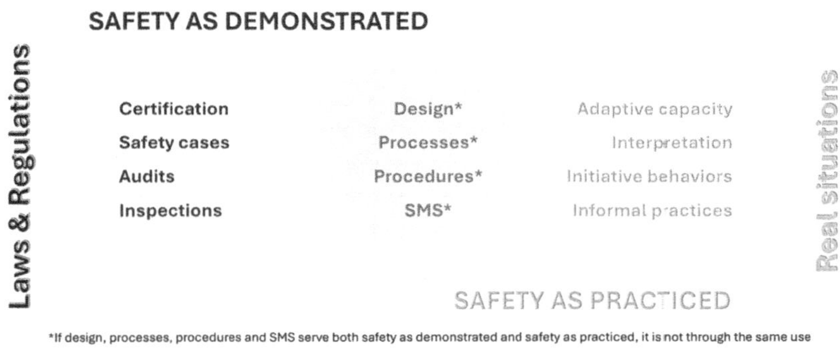

Fig. 1.1 Common bases and discrepancies between "safety as demonstrated" and "safety as practiced"

This overlap sometimes leads to inconsistencies, or discrepancies between the "as demonstrated" corpus and "as practiced" activities: Front-line worker will sometimes deviate from procedure because operational conditions are not identical to those anticipated, or because their experience tells them that an alternative approach will be more effective in these circumstances, or because they know that some parts of the official procedure were written more to manage organizational liability than for real-life feasibility. Most of the times, though, it is expected that the discrepancies will not extend beyond the local level. Thus, they do not represent a threat for the "safety as demonstrated" legitimacy.

While a substantial share of the safety improvements achieved over recent decades can be attributed to advances in technology, system design, and the regulatory and procedural frameworks that constrain system operation, these technical, prescriptive, and codified dimensions of safety do not tell the whole story. Safety also depends on the tacit knowledge accumulated over time by experienced front-line staff, which is transmitted through informal mechanisms such as storytelling, as described by Jan Hayes and Sarah Maslen in their chapter *The power of the informal and unspoken in safety practice* (Volume 2).

Somewhat surprisingly, safety stakeholders appear to live quite comfortably with the coexistence of safety as demonstrated and safety as practiced, despite the inconsistencies between them. One might have expected the "as demonstrated" system to evolve to incorporate improvements developed in the context of "as practiced," for example via negotiation between regulators and regulated entities. An administrative system that enriches itself through scientific insights and operational experience would seemingly constitute a virtuous process. However, the safety as demonstrated framework has not evolved significantly over time to integrate and reflect the growing body of knowledge on safety as practiced that never reached the public sphere and is known by safety experts only.

There are several possible reasons for this absence of change. A first one is that, given the overall level of safety and the considerable progress achieved over the past decades, existing frameworks function well enough and do not require major evolution. Another possibility, based on the principle of epistemological pluralism, is that the existence of multiple ways of thinking about safety is not necessarily a problem, even if they are conceptually incompatible, as long as each approach produces improvements without weakening the other. Bureaucratic inertia is another possible reason: administrative systems may simply be unable to keep pace with the development of new knowledge. Conceptual and methodological reasons are also frequently cited, particularly the lack of practical methods and tools to support a safety as practiced perspective. Alternatively, it may be that the gap between the two systems went largely unnoticed because they are managed by different actors who interact only marginally. Finally—and this is the explanation that we tend to favor—the institutionalization of the "good old safety model" as the right model for the governance of safety and as the socially accepted discourse on safety is likely to have added a lot to generic bureaucratic inertia. Because of the public, negotiated, formalized character of a governance model, it is even more difficult to make it evolve. An institutionalized model, by definition, is stabilized. It is frozen around a political equilibrium that few actors dare to challenge. Change is slowed down in the name of predictability and clarity. Institutionalized models evolve by quantum leaps, but very infrequently. Otherwise, only incremental change is tolerated. Whatever the underlying reasons, it now appears that the separate evolution of each framework has widened the gap between them, to the point that they have become largely inconsistent with one another.

1.2.2 Organized Hypocrisy and Organizational Facades: A Useful Lens for Understanding the Current Situation

Inconsistencies of this kind are not specific to safety; they are a well-recognized feature of organizational systems. Let us therefore reframe the two safety frameworks using different terms. Safety as demonstrated is essentially designed and operated to produce an account addressed to regulatory authorities and other external stakeholders. It can be understood as a form of "discourse," a term which we use in a broad sense to refer to the meanings that are conveyed by the bundle of written texts and oral communications addressing what safety is about, what it should be, and how it should be achieved. In contrast, safety as practiced generates actions that are largely internal to the organization. Safety as practiced appears to belong to the realm of action rather than to the realm of discourse. Though it is also supported by texts and talks, safety as practiced has a low degree of integration and a much lower degree of visibility. In other words, though accepted and even often encouraged, it is not institutionalized. Without denying that safety as demonstrated also leads to concrete actions or that safety as practiced may sometimes be articulated in discourse, it remains the case that the two frameworks serve different purposes. The persistence of discrepancies between them is therefore unsurprising. They are a form of what Brunsson called *organized hypocrisy* (Brunsson 1989, 1993), to describe the organizational practice of maintaining two co-existing strands of activity: a political one, oriented toward producing legitimacy in the eyes of stakeholders, and an economic one, focused on making the organization function in order to achieve its goals. As Brunsson (1993) explains, organized hypocrisy is not inherently problematic in democratic societies; on the contrary, it has several advantages. It allows organizations to cope with the gap between what can be said and what can realistically be done in practice. In contexts in which it is impossible to reconcile the expectations of diverse stakeholders—both external and internal—with organizational objectives and practical feasibility, organized hypocrisy enables organizations to maintain legitimacy by decoupling talk, decisions, and actions and by managing the resulting inconsistencies. "Talk," here, refers to what is officially said or written by organizations to stakeholders (e.g., press releases, company statements, safety documents provided for audit or certification). According to Brunsson (2007), it is not necessary for organizations to follow a linear or hierarchical process in which talk leads to decisions and decisions to action: Talk may not result in decisions, and actions may occur without prior decisions.

Organized hypocrisy "can be part of a deliberate strategy or simply an unintentional response" (Penklis 2024, p. 256) to situations characterized by conflicting and partly incompatible expectations. In short, it is a way for organizations to handle the diversity of stakeholder expectations while pursuing their own goals, which are sometimes only partially aligned with those expectations.

This phenomenon is not confined to the organizational level. Organized hypocrisy can also exist at the nation-state level, and even at the global level, for instance in activities that extend beyond nation-state boundaries, such as those of multinational corporations or NGOs (Penklis 2024). Within organizations, organized hypocrisy may also emerge at lower levels, such as between departments or across hierarchical levels (Abrahamson and Baumard 2008).

To support organized hypocrisy and the decoupling of talk, decisions, and action, organizations construct *organizational facades*, which they dynamically adapt to stakeholders and to changes in their environment in order to maintain legitimacy (Nystrom and Starbuck 1984; Abrahamson and Baumard 2008). These facades conceal underlying practices that, if revealed, might lead stakeholders to perceive problems and withdraw their support. In other words, organizational facades hide backstage practices and are intended to reassure.

Abrahamson and Baumard (2008) distinguish three types of facades. First, *rational facades* respond to stakeholder expectations regarding how the organization should operate. As Penklis (2024) notes, citing Crossan et al. (2005), "this facade can allow for improvising and responding to urgent demands caused by unexpected or uncertain events while still maintaining the appearance of a certain level of certainty" (p. 102). Second, *progressive facades* convey norms of progress and give the impression that organizations are using state-of-the-art (or buzzword) management techniques, even when this is not fully the case. The authors emphasize that such facades can act as incentives or levers for organizational improvement when they depict progress that exceeds reality, by projecting a desirable image and encouraging convergence toward it. Third, *reputation facades* contribute to the organization's positive image by displaying accounting and rhetorical symbols that are valued by critical stakeholders (Abrahamson and Baumard 2008).

In sum, living with organized hypocrisy—and thus with the decoupling of talk, decision, and action—is inevitable in contexts characterized by irreconcilable stakeholder views, and it enables organizations to operate in dynamic environments. The key challenge lies in managing these inconsistencies in ways that preserve legitimacy. According to Brunsson (2006), the illusion of rationality must be maintained through organized hypocrisy. However, Effron et al. (2018) show that audiences do not necessarily react negatively to perceived inconsistencies between words and deeds. What matters is whether the audience interprets the misalignment as a "claim to an unearned moral benefit" (Effron et al. 2018, p. 64). Negative reactions arise only when the inconsistency implies moral discredit, such as decisions that are publicly justified by safety concerns but interpreted as being motivated by profitability. Another situation that can make hypocrisy unmanageable is when initial false statements fail to restore trust, leading to new lies: "when lies do not result in the truth, requiring more lies and thicker facades" (Abrahamson and Baumard 2008, p. 450).

1.2.3 Organized Hypocrisy Around Safety Management and Governance

Organized hypocrisy is both inevitable and potentially beneficial, provided it is adequately managed. How, however, does this concept translate to the field of safety management? Over recent decades, substantial improvements in safety have been achieved even as talk and action have become increasingly decoupled. At the same time, major accidents such as the Boeing 737 MAX crashes have revealed organizational facades that concealed unsafe practices (EAIB, 2022). This raises a critical question: to what extent is organized hypocrisy conducive to safety, and when does it become detrimental? This section begins with an overview of safety management and governance through the lenses of organized hypocrisy and organizational facades, before reflecting on the potential safety implications of inconsistencies between talk and action.

1.2.3.1 A Systemic Hypocrisy

The safety model based on anticipation and a priori control has historically led to significant safety improvements in high-risk industries and as a result has established itself as a source of legitimacy. The partial decoupling between safety as demonstrated and safety as practiced (between talk, decisions, and actions regarding the production of safety) was not initially intentional or even perceived. Rather, it emerged gradually as a consequence of conflicting expectations between safety authorities and operational organizations: on the one hand the need for a technical and organizational framework that produces safety using mechanisms that can be analyzed, inspected and explained, and on the other hand the need to operate high-risk activities efficiently and safely despite occasional unforeseen circumstances. Though individuals at safety authorities typically recognize the need for initiative at the sharp end of operations to deal with situations that were not anticipated by the system design and formal operating procedures, it is difficult for their institutions formally to acknowledge the need for such deviations from the safety as demonstrated framework and its associated rituals of verification that provide comfort and legitimacy (Power 1997). Over time, these inconsistencies have become increasingly visible, at least to certain actors within industry and regulatory bodies.

In addition to the safety as demonstrated discourse prepared for safety authorities (and, in some industries, for internal control bodies), firms produce safety as demonstrated—or as claimed—discourses for other stakeholders, such as employees, board members, shareholders, and society at large. Different stakeholders hold divergent expectations. Without delving into excessive nuance, shareholders generally seek a level of safety that does not jeopardize profitability, neither exposing the organization to an unacceptably high risk of accidents and their financial consequences, nor requiring what are perceived as excessive and unnecessary safety expenditures. Boards of directors tend to expect a workable compromise between ensuring control and an economically "reasonable" approach to safety. Employees seek safe operations and acceptable working conditions, while also ensuring sufficient production to sustain the organization. Societal expectations are even more diverse, encompassing these concerns as well as others, such as maintaining a high level of industrial safety without harming the environment.

Within this landscape of conflicting expectations, high-risk organizations must continuously adapt or re-craft their safety discourse and the associated facades, both for individual stakeholders and for their broader stakeholder constellation, in order to maintain legitimacy. Simultaneously, they have to maintain consistency and coherence over time, as major changes in their discourse could be interpreted as an admission that safety was previously badly managed. There is little leeway for any "revolution in safety affairs."

Organized hypocrisy in safety management is not confined to high-risk organizations themselves. Safety regulators also engage in forms of organized hypocrisy to cope with conflicting and partly irreconcilable demands. Ensuring high-quality and effective regulation and oversight while maintaining independence from industrial actors has long been central to regulators' missions. However, additional expectations have emerged over recent decades, some of which are mutually contradictory. Safety authorities have been asked to extend their scope to new technologies (e.g., artificial intelligence, despite the specialized expertise required) and sometimes to new domains altogether (e.g., security), while maintaining—or even reducing—their resources, despite the growing number of organizations and issues under their remit. As a result, regulatory practices may become increasingly decoupled from the facades they present to reassure different stakeholders. For instance, Baram and Bieder (2022) show how some authorities, constrained by limited resources and slow processes, end up requiring conformity with industry-developed standards rather than developing regulatory requirements from scratch, thereby undermining the facade of regulatory independence.

Citizens, too, may decouple talk, decision, and action with respect to high-risk activities. While, as citizens, our public discourse often emphasizes safety and environmental protection, our individual behavior reveals a preference for low-cost and convenient products and services. Similarly, some activist groups advocating "safety first" prioritize issues that enhance their media visibility over those most relevant to safety outcomes.

The organized hypocrisy of each stakeholder is dynamically adjusted and, in turn, influences the organized hypocrisy of others. As illustrated by Størkersen et al. (2020) in maritime transport, the move away from "command and control" approaches to regulation by some safety authorities induced internal over-regulation within high-risk organizations. This took the form of compensatory internal processes, procedures, and additional facades aimed at offsetting the reduced specificity of regulatory requirements.

Størkersen and Dekker, in their chapter *Suffocation by management systems* (Volume 2), analyze, beyond radical deregulation that makes the headlines, the softer forms of deregulation that are implemented in almost all countries. This trend aims at reducing the regulation costs for the state or for companies, or for both. On the companies' side, however, there is strong evidence that deregulation produces an adverse effect of suffocating safety management with increasing audits, internal bureaucracy, and red tape, meaning increased, not reduced costs. More importantly, the new requirements induced by deregulation distract companies and workers from real safety issues.

Overall, safety management and governance appear to be characterized by a systemic form of organized hypocrisy, in which the various stakeholders' hypocrisies intertwine in a subtle and seemingly fragile equilibrium. A key point here is that no actor has a significant degree of control over the whole system, not even regulatory authorities. As noted before, this systemic interconnectedness is probably a factor of inertia. Inertia is two-edged: The system is stabilized, but at the cost of reduced adaptive capacity. While this balance has so far been maintained, it raises concerns about its sustainability.

A focus on managing inconsistencies directs attention toward the day-to-day evolution of safety management and governance. Repairing emerging gaps in existing facades, erecting new ones, or adjusting action to meet short-term organizational objectives fosters a perspective limited to the present and to currently salient stakeholders and scopes. In other words, the work of maintaining legitimacy tends to distract from deeper, broader, and longer-term reflection. This approach has appeared sufficient until now and has fostered a form of confidence that solutions (particularly technological ones) will emerge in time to ensure future safety.

A similar phenomenon can be observed in the context of climate change (Frantz and Mayer 2009), where experts' warnings, such as those of the Meadows report published in 1972, were largely ignored until the impacts of climate change became undeniable. This can be interpreted as a macro-level hypocrisy, in which stakeholders collectively constructed a reassuring facade about the future without engaging in immediate action to address long-term challenges. Yet the pace and magnitude of ongoing transformations may require looking further ahead and taking action today to ensure safe operations tomorrow, in a context that may differ substantially from the present one. Existing foundations may not be able to support ever-thicker facades.

1.2.3.2 Organized Hypocrisy: A Fruitful or Detrimental Approach to Safety?

As argued by Brunsson (1989), organized hypocrisy is both inevitable and capable of yielding positive effects. In the domain of safety, organizational facades make it possible to keep actual practices out of sight, thereby enabling organizations to confront delicate operational situations that require rapid adaptation or, at times, experimentation, without having to engage in a lengthy formal demonstration process. As noted by Penklis (2024), drawing on Crossan et al. (2005, p. 102), organizational facades allow organizations "to improvise and respond to urgent demands caused by unexpected or uncertain events while still maintaining the appearance of a certain level of certainty." In this sense, facades enable organizations to move beyond strict compliance with external requirements and to rely on alternative, typically less formal mechanisms that may appear improvised to stakeholders, but which are essential to maintaining safe operations.

Such practices include those emphasized in High Reliability Organization (HRO) theory, such as maintaining high levels of training for front-line staff that include aspects beyond their narrowly defined task, and delegating decision-making authority to front-line operators, with managerial support, in unanticipated situations. These practices do not fit easily within the anticipation-and-control model expected by authorities and may even appear partly contradictory to it. They may also be viewed unfavorably by other stakeholders. Shareholders, for instance, may see extensive training as an inefficient use of resources. Similarly, safety authorities and boards of directors may perceive the delegation of decision-making authority to front-line staff as a weakening of hierarchical control.

Decoupling talk and action can thus support safety provided that three conditions are met: first, that action genuinely supports safe operations; second, that facades are managed in ways that preserve stakeholders' trust in the organization, and thereby its freedom to operate safely; and third, that inconsistencies between facades and practices do not undermine the organization's legitimacy, particularly by eroding its moral credit. This situation entails a form of organizational schizophrenia, in which the organization does not fully do what it claims to do, while nevertheless believing that what it actually does is the right course of action. Crucially, this position must be sustained over time. Under these conditions, facades may function either as ideals to strive for and/or as permanent reminders of the imperative of safety, without being contradicted by accidents or major incidents. In doing so, they can provide genuine support to safe action.

However, top management may be tempted to prioritize safety as demonstrated and the maintenance of facades, since this is where immediate negative consequences are most likely to materialize, such as the loss of license to operate, reputational damage amplified by rapid media dissemination, or legal action. Moreover, managing safety as demonstrated is itself resource-intensive. As Abrahamson and Baumard (2008) argue, the continuous re-crafting of facades to respond to emerging threats to legitimacy can become a full-time occupation, requiring employees to engage constantly in decisions about how facades should evolve. This dynamic may lead to limiting or reducing the resources—time, attention, manpower—allocated to safety as practiced, either as a cost-saving strategy or as a result of insufficient understanding of what effective safety practice entails.

Safety as practiced, for its most part, constitutes a semi-clandestine system that is difficult for top management to observe and is therefore easily overlooked. Front-line operators and middle managers may attempt to compensate for decisions that fail to support—or actively undermine—safety as practiced. However, safely adapting to real operational situations is not always possible when resources, processes, and procedures create obstacles to action. Under such conditions, adaptation may give way to the normalization of individual or collective deviance. Furthermore, as facades become increasingly elaborate and disconnected from actual practices, employees may find their work less meaningful. This can have consequences not only for operational safety—for instance, through reduced reporting of deviations when employees no longer trust the organizational framework within which their safety-related activities are embedded—but also for employees' health and mental wellbeing. While such harm currently falls outside the scope of industrial safety as traditionally conceived (i.e., the prevention of major accidents), it remains harm nonetheless and may also generate indirect and longer-term effects on industrial safety.

Sustaining safety as practiced requires a combination of conditions, including the presence of competent and institutionally recognized safety experts, an organizational culture supportive of safety, a strong commitment to safety on the part of operational managers. It also implies that staff can distinguish between activities that are useful for safe operations and those mainly intended to pacify outside observers and are willing to participate in this theater without turning it against the organization, for example by exposing discrepancies to external stakeholders. These conditions, like safety as demonstrated and the inconsistencies between talk and action, must themselves be actively managed if organized hypocrisy is genuinely to support safety.

Against this backdrop, it is not obvious that the current system of organized hypocrisy in safety management and governance remains viable in the face of ongoing environmental, technological, political, and societal transformations. The following section discusses this question in the context of democratic societies.

1.3 Turbulences Ahead: The Main Challenges to the Current Organized Hypocrisy

The world and human societies are currently confronted with multiple (r)evolutions and macro-level trends. While adaptation to change is not new, including in the domain of safety, the scale and pace of contemporary transformations are unprecedented. Some of these developments operate at a global level, such as climate change, geopolitical tensions, or radical political decisions taken by countries with major international influence. Others unfold at a previously unknown speed, as illustrated by the rapid expansion of social media or of artificial intelligence systems. Many of these transformations affect the safety of high-risk activities either directly, within the traditional boundaries of industrial safety, or indirectly, through their broader ecosystems and the interactions between the two. Collectively, they call for exceptionally rapid adaptation, which may be disruptive in a field such as safety, where responses have traditionally followed a cautious and therefore relatively slow-moving trajectory.

This section examines the ways in which current and emerging transformations challenge the prevailing forms of organized hypocrisy surrounding industrial safety. It is structured around the elements that are being destabilized, drawing together the various sources of destabilization. While the phenomena discussed are often interrelated, they are addressed separately for analytical clarity. The first subsection focuses on the destabilization of societal relationships to safety, while the second examines the destabilization of the conditions under which safety management is implemented.

It should be emphasized that this diagnosis is based on the current situation and may rapidly become outdated, given the instability of the global context and the accelerating pace of change.

1.3.1 Destabilizations of Societal Relationships to Safety

1.3.1.1 A Growing Number of Priorities Competing with Safety

Catastrophic industrial accidents have become extremely rare in most contexts. At the same time, other issues such as climate change, security, and mental health have gained increasing prominence and urgency, competing with, and in some cases overtaking, safety in terms of perceived priority. These "new" challenges are becoming central concerns at the societal, industrial, and even academic levels.[2]

---

[2] However, recent years showed that priorities might be volatile. Describing climate change as "the greatest con job ever perpetrated on the world" in his speech at the 80th United Nations summit in New York in September 2023, Donald Trump illustrated the fact that talk at State level may create or dismiss priorities and can be followed by radical action (consistent with talk) such as cutting public subsidies to all climate change research relying on them.

In his chapter *Towards New Imaginaries of Risk and Security*, Jean-Christophe Le Coze (this Volume) argues that the considerable development of globalization, and the resulting level of interconnection, constitutes a "global turning point" that affects societies at their very foundations. Hence "risk" and "safety" can no longer be conceived as scale-invariant notions. He proposes distinguishing three classes of risk; each associated with a scale of organization, and constituting at each level a circular causality loop, coupled to the other two by complex links. The first loop refers to "sociotechnical risks" (industrial and infrastructural risks). The second to "systemic risks" (linked to the globalization of energy, resources, and value chains). The third to "existential risks" that threaten the habitability of the planet.

Acts of violence and malicious behavior are also increasingly reported within societies and organizations (Hansen and Antonsen 2024). It is easier than in the past to exert nuisance or disruptive power, in particular due to widespread access to digital technologies. Moreover, information about such events now circulates far more rapidly and broadly than before.[3]

This shift in perceived priorities has concrete implications. In Europe, for instance, approximately half of the academic resources previously dedicated to safety have, over recent decades, been redirected toward environmental impacts and other more fashionable topics (see Nævestad and Wennberg, Volume 2). Safety departments within high-risk organizations and regulatory bodies have experienced significant reductions in funding, human resources, and expertise (see Licu, Volume 2). This relative retreat of safety-related investment must nevertheless be interpreted in light of the dramatic improvements in safety performance achieved over the past decades in many industry sectors.

Safety facades, however, largely remain intact and sufficiently reassuring for most stakeholders: safety authorities, whose safety as demonstrated focus has not been weakened by limits to their resources; citizens, whose attention now encompasses additional concerns such as climate change and state-sponsored terrorism but who are likely to react as strongly as before in the event of a major accident; and top management, for whom formal safety management structures remain in place. To date, this confidence has not been contradicted by observable events.

In practice, however, people closer to the sharp end of operations are increasingly confronted with new and growing safety challenges arising from ongoing and future transformations (see subsequent subsections). As reassuring and well-polished facades continue to ensure that safety is largely taken for granted, there is a significant risk that safety as practiced may be neglected and that inconsistencies between talk and action will widen, increasing the vulnerability of the current system of organized hypocrisy.

---

[3] https://foncsi.org/sites/default/files/Publications/Publications_FR/FONCSI_Ateliers_Synthese_25mars2025.pdf.

1.3.1.2 The Disintegration of the Social Contract Implicitly Underpinning Traditional Safety Management and Governance

The social contract governing high-risk industries, which historically led to a safety governance model based on anticipation and control in which regulatory bodies act as representatives of societal interests, is increasingly being questioned. The relationships between industry, government, and society are undergoing significant changes that call for a rethinking of traditional safety management and governance structures (see Grote, this Volume).

Transformations in the world and society, along with accidents and scandals (such as those related to health or industrial crises), are reshaping the conditions for social acceptability. While social acceptability was once viewed primarily as a matter of informing and explaining issues to the public, it is now increasingly understood by social scientists—particularly in democratic nations—as the population's consent to a project or decision based on collective judgment that the proposed course of action is superior to known alternatives, including the status quo (Gendron 2014). This shift in understanding reflects an increasing desire among citizens to be more directly involved in decisions that affect their communities and lives, beyond merely casting a vote.

Despite the challenges inherent in such participation (Bieder et al. 2025), some countries have even mandated citizen involvement in the planning and approval of high-risk facility construction projects. However, under the current safety governance model, citizens' interests are still largely represented by regulators, and they do not play a direct role in decision-making. Furthermore, there are often conflicting views among citizens regarding high-risk activities and their safety management. To date, citizens have largely been excluded from the shaping of the external measures intended to reassure and protect them. Nevertheless, with the rise of new information and communication technologies, some groups have managed to access insider information, to make their voices heard, and at times to influence safety practices. In this new context, relying solely on the traditional authority–industry relationship for safety governance may no longer be feasible. Some groups opposed to these decisions may perceive them as part of a deceptive collaboration between authorities and industry to maintain the industry's prosperity, which could lead to intense negative reactions.

The rapid development and widespread use of new technologies, along with the variety of their applications, whether intended or unintended, pose new societal challenges. AI serves as a prime example of this phenomenon. Despite its profound influence on many aspects of daily life, including the organization of societies, businesses, political systems, justice, and health care, there appears to be no established social contract for AI systems (Snyder Caron and Gupta 2020). High-risk industries are no exception to the influence of AI, which is being applied in areas such as recruitment, industrial process control, quality control, and, crucially, safety management at both operational and strategic governance levels. The introduction of AI into safety governance in high-risk industries presents new challenges, particularly in the absence of a clear social contract for AI systems.

The chapter *How Can We Make AI Understand Safety?* by Stathis Malakis, Marc Baumgartner, Anthony Smoker, and Erwin Langejan (Volume 2) discusses the challenges involved in integrated AI-infused components in safety-critical systems, including the difficulty in interpreting model outputs and risks related to misinformation. They analyze governance methods that could help ensure that accountability is distributed fairly when these opaque technologies are introduced into complex multi-level systems while aiming for alignment with societal values.

As high-risk industries expand beyond traditionally industrialized nations, the Western model of safety governance, based on the historical social contract, is neither the only nor the dominant framework. The diversity of social contracts governing safety is further complicated by political shifts in some countries, where there is a growing trend to weaken, or even eliminate, governmental regulatory frameworks in favor of industry self-regulation. While the absence of regulation and oversight by the State does not imply the absence of safety management, the underlying principles, objectives, and reference points may differ from traditional frameworks. In certain contexts, these alternative safety management systems appear to be effective, both in terms of safety and of financial performance, even if they are still in a maturation phase.

In this evolving landscape, new facades must be developed to address the uncertainties surrounding the role of AI systems. This includes the involvement of new stakeholders, previously outside the realm of safety management and governance, whose participation introduces a range of technological, organizational, legal, and ethical complexities.

1.3.2 Destabilizations of the Operationalization of Safety

1.3.2.1 *An Increasing Mismatch Between Safety Management & Governance and Technological, Societal, Political Temporalities*

Safety management and governance have traditionally relied on conservative approaches, whereby comprehensive a priori analyses, tests, safety cases, and various forms of demonstration are conducted before any new technical system, modification, or procedure that may affect safety is implemented. Although high-risk industries have undergone many changes since their inception, and safety management and governance have adapted accordingly, the pace of certain evolutions has shifted markedly over the past decades. Extreme environmental events are more frequent; information and communication technologies have rapidly transformed the ways in which people interact; AI systems are now ubiquitous, whereas deep learning capabilities did not exist twenty years ago. More recently, radical political decisions have altered the economic and academic landscape almost overnight. This technological, political, and societal environment has accelerated the pace of change, thereby widening the gap between the scope and speed of transformations in the world and in societies, and the capacity of existing safety management and governance approaches to respond.

While ongoing and forthcoming transformations require unprecedentedly rapid adaptation, the processes associated with the institutionalized "as demonstrated" framework (which traditionally respond over long time horizons) are profoundly destabilized and unable, in their current form, to keep pace. As an illustration, following an accident, videos and podcasts containing substantial safety-related content are sometimes shared on social media within twenty-four hours, whereas official investigations and the resulting recommendations may take years to be communicated and implemented.

This time lag between organizational action and the apparent availability of insights needed to enhance safety may trigger strong reactions. Stakeholders, for instance, may interpret it as resulting from a lack of effort, or worse as an attempt cover up mistakes. Employees, in turn, may choose to rely on quicker unofficial means, tools, or approaches within organizations, thereby modifying practices in ways that are more or less visible to management, more or less safe, and ultimately amplifying the inconsistencies underpinning organizational hypocrisy.

1.3.2.2 An Increasingly Diverse and Complex Industrial Landscape

The fragmentation of organizations and the globalization of the economy have already been identified as drivers of increasing organizational complexity (Milch and Laumann 2016). In addition to these well-documented factors, more recent phenomena, operating at multiple scales, further contribute to the complexity of the industrial landscape or introduce additional sources of uncertainty, as discussed below.

a. *At the business and safety governance level*

In the context of climate change and geopolitical tensions, there is an increasing interrelationship between political decision-making and industrial activity, which manifests in several ways. Geopolitical developments raise sovereignty concerns and call into question previous industrialization choices. Certain high-risk industries are either encouraged to dismantle or compelled to expand at unprecedented rates (e.g., nuclear power in Europe and the USA), depending on their environmental impact and on the priorities of the governments in power. Recent years have illustrated the volatility of political agendas, generating significant industrial uncertainty or instability. At the same time, some large industrial corporations themselves play an active role in current tensions, including economic and political ones. In all cases, these situations and the associated sudden decisions may entail unexplored safety challenges.

Moreover, in recent years, the evolution of certain political regimes and the development of high-risk industries in some countries have led to the emergence of alternative safety governance approaches characterized by minimal, if any, governmental regulation. In these contexts, safety tends to be self-regulated by industry actors in ways that differ markedly from traditional anticipation- and control-based safety management and governance practices. This appears to be the case, for example, for autonomous vehicles in the USA and China, where certain industry players have designed relatively safe products in a profitable manner, though their occupational safety performance does not always appear to be satisfactory.

b. *At the work level*

Although the safety as demonstrated framework typically assumes that sufficient resources—both in quantity and in quality—are available to ensure safe operations, practice often requires adjustments to chronic understaffing. Until now, this situation has generally remained manageable, despite significant challenges for safety (e.g., co-activity between workers with very different professional backgrounds, cultures, or languages), despite the sometimes considerable human cost for employees, and despite the degradation of some workers' rights (e.g., platform work). However, the rapid evolution of technologies and the expansion of certain industries point to unprecedented recruitment and competence challenges, including workforce shortages in some sectors, skills that are no longer adapted to evolving work environments, and deficits in emerging technological domains (e.g., artificial intelligence) (Tillement et al., 2026).

The extent to which these competence-related issues affect high-risk activities varies across industries, countries, and regions. Education policies, labor-market integration mechanisms, and political strategies for technological development all play a role in shaping the scope of the problem. Nevertheless, existing arrangements—such as reliance on precarious employment or underqualified personnel—which, behind the facade, have allowed certain countries and industries to maintain an acceptable level of safety, may no longer be sustainable given forthcoming transformations.

> In their chapter *Health and safety challenges in precarious work,* Ulla Forseth and Hilde Bjørkhaug (this Volume) address the emergence of new precarious employment relations and how this development challenges existing safety. The chapter reflects a two-speed system in which we are deliberately continuing to protect an increasingly privileged majority of employees, at the cost of creating a marginalized but growing group of precarious and poorly protected workers.

A related issue raised by industrial representatives refers to changes in the attitudes of new generations of employees toward work, with greater importance attributed than in the past to finding meaning and purpose in their jobs, their relationship with information and social media, and the similarities between their personal and professional modes of engagement. These observations also point to a certain tendency to question or resist externally imposed orders. While a discussion of generational differences lies beyond the scope of this book, it is important to note that such changes are perceived by industrial actors and are addressed as such in organizational practices.

Another major transformation of the industrial landscape stems from the introduction of new technologies and the multiple ways in which they challenge the existing status quo in safety management and governance. By design, the behavior of systems based on technologies such as artificial intelligence evolves through use. This unprecedented characteristic of technical systems is not addressed by existing certification frameworks, which rely on the assumption that system behaviors remain stable over time and that their responses can be anticipated and demonstrated to be safe a priori. As a result, new questions of trust and responsibility arise, affecting both certification activities and operational decision-making and action.

Adopting a social interaction perspective, Petter Almklov and Stian Antonsen in their chapter *Lost in representation* (Volume 2) analyze how digitalization affects the organization of work by giving more weight to data than to interpersonal relations, disembedding social elements out of the tasks through standardization, and physically distancing organization members from one another. The authors also question the inevitability of these changes, arguing that "attempts to streamline, structure and govern social interaction from above are likely to fail also in the digital future as much as it does today."

This evolving landscape also challenges the social interactions that shape work situations and that play a significant—though often informal and largely invisible—role in safety.

Teemu Reiman in his chapter *Safety and culture in an organizational context* (Volume 2) unpacks the concept of safety culture by distinguishing between four different conceptualizations: as a remedy, as a causal explanation, as a dynamic process, and as an emergent structure. Addressing both researchers and practitioners, he calls for an integrated approach, "considering the overall system as greater than the sum of its parts."

Overall, the evolution of industrial settings and of their environment raises many issues at the workplace level that are either overlooked or addressed through solutions that are socially debated and are opposed by a part of the population (e.g., precarious employment, use of AI).

c. *At safety management level*

The traditional approach to safety management in high-risk organizations relies on a dedicated safety department intended to ensure an independent perspective on trade-offs between production and safety. This arrangement has often resulted in industrial safety being treated as disconnected from core business objectives (see Licu, this Volume). This siloed approach was further reinforced by the growing prominence of environmental and security concerns, which led high-risk industries to establish dedicated departments and, in many cases, management systems to address these increasingly pressing priorities. The outcome is the coexistence, within many high-risk organizations, of multiple structures, management systems, and practices, each focused on a specific safety objective (e.g., industrial safety, environmental safety, security, occupational safety), with limited articulation among them.[4] A similar development occurred at the level of regulatory authorities, either through the creation of distinct authorities for each objective, again with minimal coordination, or through the expansion of the mandates of existing authorities (in particular, broadening a historical industrial safety mandate to add security).

At the same time, the tight coupling between high-risk organizations and their broader ecosystems has become increasingly evident. The entanglement of politics and industry has been illustrated earlier, and the growing frequency of Natech accidents (technological accidents triggered by natural events) further underscores this trend. Numerous studies and accident analyses have also highlighted the interrelations between safety and security. In simple terms, an approach to safety that is addressed in silos—and even in multiple silos, one for each "type of safety"—already reveals its limitations. While the gradual emergence of these concerns may explain the initial development of parallel organizational structures, maintaining poorly interconnected silos is becoming an indefensible facade and may lead to inefficient practices in a world that does not recognize these somewhat artificial boundaries.

At the level of safety management practices, the rapid development of technologies such as big data and deep learning offers new opportunities[5] for continuous monitoring, large-scale analysis of operational conditions and practices, and the identification of safety concerns. This data-driven approach redirects attention toward the operational level, whereas managerial discourse on safety has largely shifted since the 1990s toward organizational factors and safety culture. Whether this shift reflects genuine technological blindness or can be interpreted as a "moral crumple zone" (a deliberate strategy implemented to deflect responsibility away from managers and executives) may vary across contexts and audiences.

---

[4] Management systems often follow a harmonized structure, following the influence of ISO, but the organizational level which is empowered to arbitrate between competing priorities and the associated actions is generally very remote from the sharp end, where the conflicts appear in practice.

[5] The introduction of these technologies in safety management and governance practices also raises a number of questions (e.g., responsibility, privacy) and can create a partial blindness by narrowing the prism through which safety is managed, focusing the organization's attention only on the aspects that be captured by data. See Foncsi, 2025, for a comprehensive analysis of the AI-safety issue.

1.3.2.3 A Generic Approach to Safety Management and Governance but an Increasing Variety of Contexts

Whether at the level of authorities or within high-risk organizations, the documentary framework underpinning safety management and governance is typically designed to be sufficiently generic to apply across all intended contexts. For example, the ICAO Annex 19 standard concerning safety management in aviation applies uniformly to all aviation activities (e.g., airlines, airports, aircraft manufacturers, air navigation service providers, training organizations) worldwide, without distinction based on local culture, operational environment, or political context, and regardless of organizational characteristics such as size or business model. Similarly, within large international high-risk organizations, internal rules, processes, and procedures tend to be developed in a centralized and generic manner, despite the diversity of the contexts in which they are implemented.

As noted earlier, the increasing diversity and complexity of the industrial landscape, combined with the globalization of high-risk activities, multiply the range of contexts to which these generic safety management and governance frameworks are applied. The impacts of climate change, for instance, vary from region to region, and specific phenomena may require local responses. Likewise, the human, technological, and economic resources available to operate high-risk facilities vary significantly across regions. Efforts to encompass this diversity have led to the proliferation and growing complexity of reference documents and rules. This trend, combined with the multiplication of documents aimed at expanding the scope of regulated aspects, has produced an increasingly complex documentary framework that reveals several limitations. These limitations extend beyond the "classical" need to adapt rules, processes, and procedures to real-time situations, adaptations that have long been part of what has been described as a form of collective hypocrisy (Le Coze & Journé 2024a).

One limitation concerns the applicability of rules and reference frameworks. While large and mature organizations can mobilize the resources required to read, understand, interpret, and translate extensive documentary frameworks, many smaller organizations cannot afford such efforts. This challenge is compounded by the fact that the development of rules or sectoral standards is largely driven by large companies acting as representatives of the industry as a whole.

Another limitation relates to the pace at which reference documents are produced and revised, compared with the speed of change in the broader technological, environmental, political, social, and economic context, as well as in front-line practices. The issue of regulatory adaptation does not affect all high-risk domains in the same way. Sectors such as railways, for example, operate infrastructures and systems with lifespans of several decades and rely on rules and regulations that are designed to be stable. As a result, the documentary framework exhibits significant inertia, even though railways are tightly coupled with environments and societies that evolve rapidly. In contrast, sectors with a shorter history of institutionalized safety management and governance, such as construction or health care, tend to demonstrate greater agility in adapting rules.

However, the ability to adapt rules and procedures to a changing context is not sufficient in itself. The relationship to rules varies across domains. In some industries, such as construction, the balance between compliance and initiative is not driven solely by the need to adapt rules to specific situations, but also by the level of knowledge of the rules and the willingness to apply them. The context of co-activity—often involving precarious workers, employees from multiple organizations, and diverse cultural backgrounds, as noted earlier—further complicates this dynamic.

Analyzing three recent examples of conflict between regulator and regulated organization, Wiig and Calderwood (this Volume), in their chapter *Risk regulation at the crossroad*, argue that future regulatory practice needs to look beyond individual failures and promote a system perspective placing trust at the center of the principles of risk regulation.

The notion of standardization has long underpinned the safety facades of high-risk activities. In inherently international sectors such as aviation, it has served to promote the idea that safety levels are equivalent regardless of where one flies or whether the airline is large or small, legacy, or low-cost, through the application of high-level standards and regulatory requirements. A similar logic applies to multinational companies operating high-risk facilities worldwide, where safety facades display ostensibly uniform high-level standards. However, as ongoing and forthcoming changes continue to diversify contexts, whether in terms of organizational characteristics, ecosystem features, or interactions between organizations and their environments, maintaining partially common facades while actual practices diverge across contexts could lead to situations of untenable hypocrisy and may ultimately erode the trust of certain stakeholders.

1.3.2.4 Citizens' Involvement in Safety Practices

In addition to the evolution of society's relationship with safety, which was addressed in a previous section, recent developments have seen citizens increasingly play more direct roles in the implementation of safety measures. This participation can take place either through their active involvement in safety practices or through the influence of their actions on how safety is managed.

Recent significant accidents and crises which had consequences extending beyond industrial sites, such as the 2019 Lubrizol disaster in France, have highlighted the fact that citizens can take on active roles in emergency situations. Their unique knowledge and competencies, such as familiarity with the local environment and customs, can be invaluable in mitigating the impacts of industrial accidents. Whether through contributing to the development of emergency response plans or by directly participating in the response to an incident, citizens serve as a crucial resource, particularly when traditional institutions or organizations responsible for safety management face resource limitations. Notably, the use of social media has proven particularly effective in these contexts, increasingly facilitating communication with local populations during emergencies.

However, this same technology plays a complex role at the interface between the public and high-risk organizations in other contexts, sometimes with more worrying implications for safety. The public now has unprecedented access to information—whether accurate or false—about events within organizations, including incidents. The dissemination of such information can create an environment of "social blaming" and significantly affect the reputation of an organization. This can influence internal safety practices, either positively, by reinforcing the importance of safety when it might have been insufficient, or negatively. In some cases, the pressure generated by media scrutiny can lead to organizational silence, with staff fearing reprisal for reporting issues, thereby creating a state of blindness within the organization.

Regardless of the specific outcome, citizens are playing an increasingly important role in safety practices, no longer limited to merely observing or influencing the decisions and talk built into the facades. In the past, citizens were often seen as passive spectators, with limited access to first-hand information about an organization's actions. Today, some are very well informed of what happens behind the facade and are able to reach other stakeholders: retired workers who share their insider knowledge on social networks, families of hospital patients who publicize the deviations from standards of care that they encounter, for example. The talk and decisions that have sometimes been used primarily as a form of stakeholder management (Brunsson 2007) may today need to be complemented or even replaced by more tangible steps.

1.3.2.5 Shifting Responsibilities and Blurred Liabilities

The intersection of accountability, liability, and safety has been a subject of study for decades (Hale and Borys 2017; Grignon et al. 2025). Liability concerns often drive the creation of extensive documentation within organizations, especially defensive documentation, or "extensive paperwork [created by organizations] primarily in an attempt to protect themselves from legal liability, rather than in the interests of safety improvement." (Lim et al. 2025, p. 12).

The evolving industrial landscape, particularly the increasing number and diversification of actors involved in high-risk activities, has further complicated the accountability and liability framework. The boundaries of organizations, as well as the regulatory frameworks that govern them, are more blurred than ever. For example, the rise of digitalization and, more recently, the increasing capabilities of AI systems, have led to the involvement of new players (algorithm developers, data providers, training dataset creators) with organizational cultures closer to the "move fast and break things" motto of Silicon Valley than to the prudent conservatism of firms that develop safety-critical systems. These firms may be based in different countries, each with its own legal frameworks and regulatory standards.

Moreover, with the advent of new technologies like deep learning, the responses of systems can no longer be fully anticipated in advance. This fundamentally challenges the philosophy underlying the establishment of clear-cut responsibilities, which traditionally relied on a predefined standard of reference.

As digital technologies continue to evolve, responsibility for safety is increasingly shifting from front-line workers to the developers and designers of the technologies integrated into safety-critical systems. These new technology actors are becoming more powerful and influential, even in the development of safety-related regulations. This shift could eventually disrupt the current safety-liability status quo, necessitating the establishment of new frameworks to account for these emerging actors. These frameworks may involve redefined roles and power dynamics within the safety stakeholder arena, as traditional standards for establishing responsibility become increasingly difficult to apply.

Engen and Lindoe, in their chapter *AI and Offshore Safety Management* (Volume 2), analyze the implications of the current and rapid spread of AI tools in the Norwegian offshore industry. Risk assessments become dominated by data security, privacy, economic, and commercial considerations, rather than by industrial and work safety. The transparency and deliberation that traditionally sustain the Norwegian Offshore Regime appear threatened by institutionalized secrecy. They argue that this example strongly suggests that AI represents more than a technological shift; it signifies a transformation in governance, culture, and power.

We can already observe such shifts in industry sectors and jurisdictions where regulatory oversight is minimal or almost non-existent, yet legal systems still attempt to assign liability after incidents. Examples include the regulation of autonomous vehicles in the USA and the off-label use of commercial chatbots for psychological counseling.

New technologies and especially AI are spreading and penetrating all work environments. High-risk industries are no exception. Through the case of aviation, Federico Franchina, in his chapter *Artificial Intelligence in the Skies* (Volume 2), dives into the detailed challenges that the introduction of AI and Europe's AI Act meant to regulate it poses to the articulation of culture, regulation and liability, and eventually to safety.

1.3.3 Unveiling Implicit Fundamental Assumptions and Unthoughts of the Current Status Quo

The ongoing and forthcoming changes in the world challenge several fundamental, often implicit, assumptions that underlie the current safety management and governance framework. One central assumption is the relative stability of the world, not only outside high-risk organizations, but within them as well. In this framework, many aspects were so deeply ingrained that they were rarely questioned or even acknowledged as assumptions. For instance, it was taken for granted that safety would be prioritized and would receive the necessary resources. Similarly, the notion that the boundaries of organizations managing high-risk systems could be easily defined seemed self-evident. The governance model, where citizens are represented by regulatory authorities, was likewise considered natural. Even the idea that a technical system could maintain stable operations without itself evolving was largely unchallenged.

However, the evolving landscape now makes it impossible to ignore the strong fundamental assumptions and limitations that underlie the "as demonstrated" framework. These assumptions are now being challenged, necessitating new ways of thinking about the future of safety management and governance.

1.4 Possible Ways Forward for Safety Management and Governance: New Hypocrisies to Live With?

The destabilizing forces affecting the current hypocrisy in safety management and governance are diverse, varying in nature, scope, timing, and scale. Facades that once seemed secure are cracking as new safety stakeholders emerge, and some stakeholders transition from passive observers to more direct participants in organizational processes, particularly during crisis situations. While facades were previously re-crafted and adapted to address evolving threats to legitimacy, and practices were dynamically adjusted to maintain the appearance of consistency, the unprecedented scale, speed, and breadth of ongoing and future transformations may render it increasingly difficult—if not impossible—to manage this situation through marginal adaptations. Stakeholder backlash could manifest in various forms, such as a loss of public trust and strong reactions, insurers withdrawing their coverage due to higher accident frequency, legal complications, and employee burnout.

In response to this unprecedented situation, several safety management and governance strategies can be considered, depending on how one assesses the magnitude of the threats to the current system. These strategies range from less ambitious, incremental changes to more transformative approaches, each requiring different levels of reasoning and action and offering varying impacts on safety (Fig. 1.2). The most straightforward and least challenging strategy, called continuity, would be to continue managing the organized hypocrisy within the existing framework. This

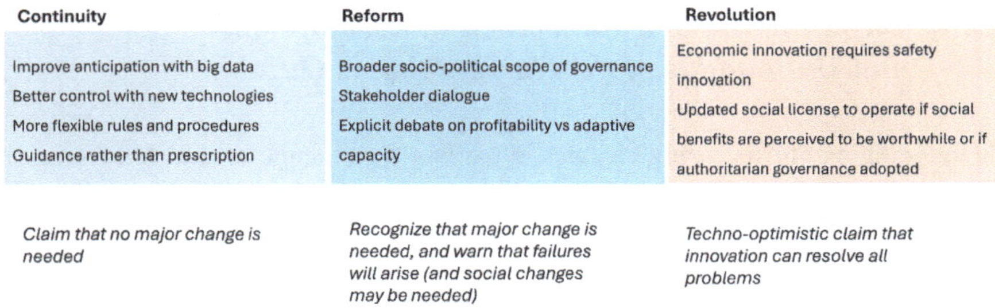

Fig. 1.2 Three possible strategies for the future of safety management and governance

approach would involve making marginal adaptations, reducing some inconsistencies while managing the remaining ones. A more ambitious strategy, called reform, would involve expanding beyond the current scope of safety management and governance to reassess the foundational assumptions that underlie today's hypocrisy and build new frameworks more suited to the ongoing transformations. Lastly, a more radical strategy which we call revolution would set aside safety concerns in favor of innovation and global societal contribution ones, thereby eliminating the very need for organized hypocrisy in safety management.

These strategies are discussed in greater detail in the following subsections.

1.4.1 Continuity: Continue to Manage the Organized Hypocrisy Within the Current Framework

The growing disconnect between safety rhetoric and action can be understood as a gradual drift that occurred over time. Initially, safety discourse promoted a model centered on anticipation and control, which was translated into procedures, processes, and rules that guided action, even though they were occasionally insufficient. Over time, situations arose that were initially unnoticed but gradually became more visible, ultimately contributing to the current "crisis" of the model. These situations now require organized hypocrisy. One way to manage this "crisis" would be to minimize the misalignment between talk and action as much as possible, while managing the remaining inconsistencies, making them easier to address.

This strategy represents a continuation of the current safety management and governance approach, focusing primarily on refining and enhancing the compliance-based and initiative-based approaches to safety. It acknowledges that the magnitude of variations around a nominal situation (possibly evolving itself over time) might be bigger than before, but all this can be anticipated.

It is built around two main objectives. The first aims to reduce uncertainties by strengthening the ability to anticipate risks and supporting the operational level in assessing and managing real-time situations, particularly through the use of technology such as large neural networks. These measures can be implemented directly by a high-risk organization. The second objective involves reassessing the level of detail and the philosophy that underlies procedures, with the goal of better accommodating the diversity of contexts encountered in practice. This would provide more flexibility at the operational level to adapt to varying circumstances.

Such an approach has implications at multiple levels. On the one hand, it requires the development, maintenance, and recognition of the competencies and capabilities needed at the operational level to effectively interpret and apply guidance materials. On the other hand, it raises important questions regarding responsibility, particularly concerning who is accountable for the decisions made at the sharp end of operations. These considerations may necessitate broader discussions at the industry level and potentially involve regulatory authorities and other relevant stakeholders. This conservative, incremental strategy is likely to gather political support from most sides, because it does not threaten established roles and positions. Whether it will be efficient in the long term is the key question.

The following subsections explore possible approaches along these lines.

1.4.1.1 Fill Gaps in the Current "Facades"

To adapt to ongoing and future transformations, one option is to improve the anticipation-and-control activities that underpin the current facades. New technologies such as big data and AI offer significant opportunities to improve in these two areas. AI systems can help predict more accident scenarios or refine the calculation of probabilities, enriching existing anticipatory methods. Some emerging risks can be anticipated through AI, which can process far more data than human experts could achieve (e.g., anticipating and mitigating Natech risks[6]). Once previously unconsidered scenarios are identified, proactive measures can be developed and implemented to manage these risks.

Additionally, new technologies can enhance control capabilities by processing a broader range of data beyond incidents and accidents, to include normal operations on a larger scale. This allows for the early identification of deviations from established requirements, potentially preventing critical events before they occur.

Challenges and feasibility conditions

Such advancements can be implemented at the organizational level, but challenges related to resource availability, both technological and human, must be addressed to improve anticipation, control, and management capabilities.

Impact on safety

As discussed earlier, expanding the coverage of anticipated situations and, consequently, the associated rules, processes, and procedures (i.e., requirements) enables the implementation of risk control measures for previously unconsidered hazards. However, being prepared for anticipated risks does not fully address the complexities

---

[6] The plausibility of these scenarios might still require human expertise to be confirmed.

of daily operations, which often involve unanticipated dynamics (e.g., as discussed in the literature from the French-speaking ergonomics school on the difference between "task" and "activity," or more recently, between "work as imagined" and "work as done"). Another concern is the proliferation of requirements in an already complex documentary landscape. AI could assist users in navigating this growing body of documents and procedures, helping to identify the relevant ones for specific situations, but the fundamental challenge remains. Furthermore, with increasing complexity and emergent phenomena, unanticipated situations are likely to become more frequent, even if available computational capabilities increase.

1.4.1.2 Adapt the Level of the Facades to Suit the Context and the Stakeholders

One underlying reason for the inconsistencies between facades and actual practices is the generic nature of current requirements, whether regulatory or internal (especially in multinational or multi-site organizations). As mentioned earlier, presenting a unified facade—one that claims everything is under control and is achieved in the same way across all levels—may reassure external stakeholders, such as the public or shareholders (i.e., the "spectator stakeholders" described by Brunsson 2007). However, this approach increases the gap between rhetoric and reality at the local level. Moreover, maintaining a uniform generic facade for employees, who are insiders and active participants in operations, may undermine credibility, as they are fully aware of the local context and the extent to which the facade diverges from actual practices.

While some inconsistencies can be addressed with additional resources (e.g., recruitment, training, tools, and equipment), others are deeply rooted in environmental factors, such as natural phenomena or socio-political contexts. Furthermore, if a safety-related event occurs in one location, the facade loses credibility across the entire organization. Adapting the facade to local contexts and specific stakeholder needs might help mitigate this issue, but it could be unacceptable to certain stakeholders, particularly in international or multi-site organizations. Ultimately, balancing generic and contextualized requirements remains a complex challenge. Attempts at contextualization do exist, for instance in aviation, where simplified requirements and compliance methods are provided for non-commercial operators of very light aircraft.

An alternative approach is to establish high-level requirements, allowing greater flexibility in their implementation at the organizational or unit level. However, as highlighted by recent studies (Størkersen et al. 2020), this often leads organizations to develop their own detailed requirements to ensure control—requirements that, in turn, serve as facades and guide actions. As with any detailed requirements, these internal rules require adaptations to the actual situation, introducing further inconsistencies. Further analysis of the underlying reasons for this reliance on internal requirements and their impact could open new avenues for understanding.

Challenges and feasibility conditions

Standardization and universality (e.g., uniform "life-saving rules" or best practices) are often viewed as symbols of strength, power, and unity. Contextualizing these requirements may undermine these symbols. Moreover, determining the appropriate level of detail and specificity for requirements and associated facades cannot be done in isolation at the organizational level, as it must consider the broader regulatory framework, which may or may not allow for contextualization.

Impact on safety

Regardless of the level of contextualization, it seems difficult to avoid detailed requirements for operational actions. Contextualizing these requirements may make them more suitable to the specific situations faced by operators, but it does not resolve the issue of unanticipated events that are not covered by existing procedures or processes.

1.4.1.3 Facilitate Compliance with Rules

Compliance with the underlying principles of the anticipation-and-control model varies considerably across different industries and organizations. While social science literature demonstrates that strict adherence to detailed requirements is neither entirely feasible nor sufficient in all contexts, neglecting applicable rules—when they are relevant—does not result in optimal safety outcomes. This is evidenced by the substantial improvements in safety achieved through regulatory and procedural frameworks established several decades ago.

The full integration of existing safety knowledge remains incomplete in many high-risk organizations worldwide. For organizations that are still in the early stages of safety management, addressing even basic safety concerns, and making efforts to anticipate and better control risks, would lead to noticeable safety improvements. Traditional safety management systems are effective even in times of complex uncertainties and continue to address a broad spectrum of safety issues. This is particularly evident in organizations where even fundamental safety approaches are not yet implemented.

Compliance with rules, processes, and procedures—when relevant—can be facilitated by emerging technologies such as artificial intelligence. These technologies, for example, can assist operators and managers in navigating complex regulatory frameworks more efficiently. However, this comes with the risk of diminishing the sense of purpose and professional expertise within the organization.

Challenges and feasibility conditions

This strategy aims to enhance safety at a broader level, specifically by improving safety outcomes across high-risk organizations on average, rather than within any one particular organization. It primarily targets those organizations that are starting from a minimal or non-existent safety baseline or where even basic compliance has not been achieved.

The central condition for the applicability of this compliance-based approach is the assumption that the premises underpinning it are valid. Specifically, this assumes a governance model in which safety authorities have sufficient resources to develop and enforce a regulatory framework. The same assumption applies to the development of processes, procedures, and internal control within organizations. However, this traditional safety model is not overly resource-intensive, either qualitatively or quantitatively.

Impact on safety

While traditional safety management and governance have historically proven effective in improving safety, significant questions remain about the continued viability of this approach. Can it still be effectively implemented with the available resources? Can it continue to drive safety improvements in organizations that have made minimal efforts in this area? Furthermore, how much longer can this model contribute to safety enhancement in an increasingly uncertain environment, and at what cost to individuals who must constantly adapt and compensate for the limitations and simplifications inherent in the existing rules, processes, and procedures?

1.4.1.4 Revisit the Status of Facades with Respect to Action: From Strict Compliance to Guidance

One avenue to manage the inconsistencies between talk and action consists in maintaining reassuring anticipation-and-control facades but remaining at a general level and translating them into requirements (rules, processes, and procedures) whose spirit and status would be revisited. Indeed, acknowledging that we need and will even more in the future need to live with uncertainty, the status of requirements, processes, and procedures needs to become closer to guidance providing support than to an absolute reference requiring strict compliance. At the company level, this means recognizing the value of initiatives taken by front-line staff, even when they consist of deviations from preplanned procedures, and implementing agile mechanisms (derogation procedures, flexible communication with the management line and with backstage experts) to support such initiatives without abandoning the traditional compliance framework. It also means that safety professionals need to develop new skills that help them navigate complex and uncertain landscapes that are subjected to rapid change (see box below). At the level of the regulator and the safety inspectorate, it may involve official recognition of the need for "test and learn" approaches to new problems and approval of cautious experimentation (Griffin, 2026). At the level of the judiciary, it requires a move away from post-accident investigations that immediately classify any deviation from standard procedure as a criminal fault that demonstrates the culpability of the people taking decisions and operating the system before the accident (Grignon et al. 2025). This would help understand to what extent and under what conditions they could be transposed to other contexts.

The changing industrial landscape calls for professionals with the skills to navigate rapid change in both technology and working arrangements while still making the best safety choices. In addition to a high level of technical competence, safety decision-makers in complex hazardous industries need strong capabilities in a range of non-technical areas to make the best choices. These non-technical skills, or "capabilities" as they are called by Sarah Maslen and Jan Hayes in the chapter titled *Supporting excellence in professional practice in a changing industrial landscape* (Volume 2), include the ability to use foresight-type strategies to make decisions in the presence of uncertainty, systems thinking abilities, an ability to pull in expertise from professional networks, and the ability to stand up for safety and the responsibilities of high-hazard firms when taking high-stakes decisions.

A shift in the philosophy or requirements underlying safety management would make the inconsistencies between organizational facades—particularly those perceived by external stakeholders and those experienced by those directly involved within the organization—more nuanced. This could potentially reduce, or at least obscure, the discrepancies between facades and actual practices, thus diminishing the risk of moral discredit for the organization in the event of an accident.

In a similar vein, while some regulators maintain a facade of independence, conducting neutral assessments for public consumption, others explore a role that blends control with coaching. These regulators aim to be more constructive than the traditional auditors who issue a simple pass or fail judgment. They assist regulated organizations in enhancing their safety practices and adopt a more nuanced approach to compliance, emphasizing sense-making (Le Coze and Journé 2024b). Although such a role might be viewed as collusion in models that deny the inevitability of hypocrisy, it could be further developed and refined within frameworks that acknowledge the necessity of hypocrisy while managing the risks of potential collusion.

In the absence of this evolution in the safety as demonstrated framework (whether through facades, practices, or both), two approaches could be considered to catalyze change. One approach assumes that both regulators and regulated organizations will recognize the tensions inherent in the current safety demonstration framework, prompting them to seek change. The challenge here lies in effectively communicating these insights from academia to practitioners and regulators. Alternative safety theories, such as those associated with High Reliability Organizations (HRO) or Resilience Engineering, are largely overlooked by most safety stakeholders. Scientific concepts, theories, and models on safety seldom percolate beyond a small academic audience. Conveying these ideas to those outside the scientific community may require new forms of communication to increase awareness of emerging challenges to safety and to promote a forward-thinking approach.

Another complementary strategy involves bridging the gap between researchers and practitioners by rethinking the positioning, methods, and outcomes of safety management research. Practitioners, both industry professionals and regulators, typically seek practical tools that can be easily implemented with limited resources. While tools are available for the current dominant model underlying most organizational facades, they are lacking for alternative models. One possible approach is to develop and provide operational tools that industries and regulators can use to assess safety practices, regardless of whether they adapt their facades accordingly.

Challenges and feasibility conditions

As previously noted, revisiting the status of safety requirements has broad implications that extend beyond individual organizations and industries, particularly in relation to the roles of authorities and issues of liability. It requires a re-evaluation of the roles and responsibilities of all the players involved, from rule makers to front-line operators, as well as of the competencies required at each level. The concept of a "regulatory sandbox," which allows for controlled experimentation at a limited scale, could help explore the feasibility and implications of shifting traditional requirements from rigid rules to more flexible guidance aimed at supporting action.

Impact on safety

From a safety perspective, revisiting the status of "requirements" and deferring to expertise when anticipation proves ineffective could lead to significant progress, as suggested by HRO theory. This approach fosters greater flexibility and adaptive capacity in response to changing demands and environments. Furthermore, it helps mitigate the "no rule, no use" syndrome described by Fucks and Dien (2017), where operators refrain from taking action—despite it being safer to do so—when no formal procedure prescribes the action, fearing potential liability consequences.

1.4.2 Reform: Build New Grounds to Keep the Organized Hypocrisy Manageable, Including in the Future

Merely revisiting or slightly adjusting the terms of the current safety-related hypocrisy may not be sufficient to address the ongoing and forthcoming transformations. One can fear that growing inconsistencies may become untenable if left unaddressed and could lead to crises, such as an increase in accidents or growing suspicion toward high-risk organizations. In turn, this could result in the loss of stakeholder support or provoke stronger reactions against these organizations. To prevent this scenario, a more ambitious redesign, implying a more extensive reframing of the current model of hypocrisy, might be required.

This reform strategy entails revisiting the foundations of the current organized hypocrisy. Acknowledging in the discourse that uncertainties and complexities are increasing and possibly leading to unforeseeable emergences, that is, situations that cannot be considered as mere variations around a nominal one, directly challenges several foundational assumptions of the existing organized hypocrisy framework.

One such assumption is the possibility of total control based on anticipation. From this perspective, anticipation alone is no longer a satisfactory solution. However, adopting such a discourse, more in line with increasingly observed situations, marks a break with the historical total control talk about safety and the underlying rational-legal framework. To compensate for this break from an apparently strong and tangible model underpinning safety discourse, a new social apparatus is needed. On the action side, embracing this perspective on the limitations of anticipation entails developing adaptive capacity and articulate it with traditional safety approaches to reach beyond their limits and live with a more complex, uncertain, and unstable world. This notably involves introducing safety margins and diversity, both technical and organizational, which are characteristics that are largely incompatible with prevailing conceptions of organizational performance focused primarily on profitability. Revisiting how the performance of high-risk organizations is defined, and how diverse dimensions such as financial performance, sustainability, industrial safety, security, environmental protection, and occupational safety are articulated, requires societal-level debates in a globalized and competitive world.

The following subsections outline the foundational aspects that need to be revisited.

1.4.2.1 Interrogating the Social Contracts at Play

Compliance with laws and regulations has traditionally been regarded as fulfilling the social contract between high-risk industries and society. This model of anticipation and control has formed the basis of high-risk industries' facades toward external stakeholders. While societal expectations regarding technology may vary across regions, adapting the current system to the aforementioned challenges requires to revisit the foundational assumptions of the historical social contract regarding safety.

> In her chapter *Toward a new social contract for safety,* Gudela Grote (this Volume) addresses growing attacks on the social contract for safety, which is normally based on a promise of full risk control through technology and business interests being kept in check by safety concerns. Among the new risks are new geopolitical realities and the advent of AI. The author proposes solutions but also acknowledges the difficulty of implementing them.

Fundamental questions that were once taken for granted must now be reconsidered: Are anticipation and control sufficient as safety strategies in an evolving world? Do societies still possess the means to achieve their safety ambitions? Are current safety management and governance practices potentially going beyond what is necessary in some instances? Reopening these discussions requires the active involvement of all stakeholders, which presents significant challenges (Bieder et al. 2025), as addressed in a later subsection on citizen participation.

Challenges and feasibility conditions

The existing social contract(s) align with broader contexts, particularly the legal frameworks in place. Revisiting the terms of this social contract, while necessary, brings about challenges in other areas, especially within the legal system.

Impact on safety

Revisiting the social contract could have profound effects on safety, although the specific outcomes (positive or negative) are not fully explored and may depend on the specific context. Nevertheless, such a shift would undoubtedly alter the ways in which safety is managed, possibly through approaches such as trial and error, or by adopting a more precautionary stance that could make innovation significantly more challenging in a globalized world.

1.4.2.2 Interrogating Companies' Economic and Governance Model

Today, most companies operate within a competitive economic model that drives most performance indicators, or at least those deemed essential, influencing most decision-making processes. In this framework, other organizational concerns, such as safety, are often viewed as constraints on short-term financial performance and, by extension, on the success of the company. This focus results in the reduction of operational margins that do not directly contribute to financial objectives. However, risk and safety management theories developed over recent decades, such as HRO theory and Resilience Engineering, are based on the premise that uncertainty cannot be entirely eliminated and must therefore be managed. These theories emphasize the importance of maintaining margins to ensure an organization's ability to adapt.

Acknowledging future uncertainties, engaging in discussions with stakeholders, and simultaneously eliminating the margins necessary to safely manage these uncertainties could, over time, erode stakeholder trust, making it difficult for organizations to maintain their facades. This does not necessarily mean that organizational practices and facades, particularly in the economic and governance domains, must align perfectly, or even that alignment is always feasible. However, some degree of change is necessary for organizations to remain credible and trustworthy in the eyes of their stakeholders.

Challenges and feasibility conditions

To drive meaningful change in organizational practices, the issue must first be brought to the attention of CEOs, shareholders, and political actors. However, the global scale of economic competition presents a significant challenge. Being the first to initiate real changes in the priorities of an organization, industry, or even an entire country is a high-stakes endeavor, particularly when the motivations for such changes are perceived to be long-term or geographically distant, as in the case of climate change.

The potential safety impacts of phenomena related to global warming, for example, can often be disregarded in favor of short-term profitability.

Impact on safety

As noted earlier, there is broad consensus that maintaining margins is essential for safely managing unanticipated situations. With the increasing uncertainties outlined previously, such situations are likely to occur more frequently. If a company's economic and governance model is reduced to mere facades, the operational margins that are not immediately necessary for economic performance—but which are critical for safety—may not be available. This would increase the company's exposure to accidents due to a lack of dynamic adaptation capabilities. Actual changes in a company's performance drivers, aimed at restoring margins necessary from a safety perspective and building adaptive capacity, could help prevent accidents even as exposure to unanticipated situations increases.

1.4.2.3 Extending the Scope of Stakeholders Directly Involved in Safety Management and Governance

New systemic risks challenge current approaches to the management and governance of safety. At present, safety management and governance remain largely in the hands of industrial organizations and regulatory bodies. However, as illustrated earlier, the role played by citizens is evolving rapidly, whether or not this evolution is discussed and agreed upon with traditional safety actors. Maintaining acceptable safety facades may be better achieved by expanding the perimeter of stakeholders who actively participate in the governance of high-risk activities. This is already the case for citizens in Switzerland, for example, who are consulted on trade-offs and where decisions respect the outcomes of popular votes. Other stakeholders could also be more directly involved, such as employees, shareholders, unions, or policymakers. While such a participatory approach to governance does not resolve conflicts between stakeholders' perspectives nor eliminate inconsistencies between discourse and action, it can provide meaning and legitimacy to decisions made, including painful trade-offs that, if announced bluntly, might otherwise trigger strong opposition (Bieder et al. 2025).

Challenges and feasibility conditions

Engaging in dialogue with stakeholders requires dedicated time, appropriate methods, and specific competencies that are not widely available today. Developing these capabilities and clarifying the mandates under which such initiatives can be undertaken remain largely to be done.

Safety impact

In the future, decisions that are painful for at least some stakeholders may be necessary in order to live safely within evolving contexts. For example, revisiting economic models may be painful for shareholders; defining new social contracts that entail a partial reduction in the benefits derived from high-risk industries in order to maintain safety may be painful for the public; and revisiting the philosophy and scope of regulation may be painful for regulatory bodies. Involving a broader range of stakeholders in discussions and decision-making may help support such difficult or radical decisions and make them more acceptable than if they were made behind closed doors and only inadvertently revealed.

1.4.2.4 Articulating Safety with Other Stakes Beyond the Traditional View on the Boundaries of Safety Management and Governance

With the increasing complexity of contexts discussed earlier, characterized by more systemic phenomena and growing uncertainties, managing or governing safety in isolation no longer appears realistic, whether as a facade or in terms of actual practices. Natech risks provide an obvious illustration of this point, as do harmful events triggered by security events (e.g., the 9/11 attacks or the bombing of nuclear facilities). Yet, despite these interdependencies, risks affecting people, the environment, and facilities and equipment are still generally addressed separately. They rely on distinct and often uncoordinated management systems and tools, implemented by organizational units that operate in silos (e.g., safety departments, security departments, environmental departments, or occupational safety departments). As discussed by Cho et al. (2015), these departments also serve a political function: They respond to specific stakeholder management requirements and are placed on organizational display. However, even as facades, such distinct entities may lose credibility as activities and stakes become increasingly interconnected. Citizens, for example, tend to think in terms of human, societal, or environmental harm, rather than in terms of the specific cause of an event or the department formally in charge of it. Maintaining acceptable hypocrisy may therefore require articulating the different organizational stakes, at least at the level of facades, while acting in ways that keep the inconsistencies with these facades apparently honest.

Challenges and feasibility conditions

Today, regulatory bodies are at least partly distinct for each type of stake (e.g., major accident hazards, occupational safety, environmental protection, security), each with its own expectations. This situation leads organizations to develop specific facades for each aspect, including dedicated internal divisions or departments. Organizational work in silos thus mirrors, to some extent, the structure of stakeholders and their expectations. Internally articulating the various stakes would therefore require articulating the expectations of the different stakeholders, which cannot be achieved at the level of a single high-risk organization alone.

Safety impact

The safety impact of articulating the various stakes together would largely depend on how this articulation is achieved. Outcomes could range from a loss of focus on safety, if it is integrated and assimilated into other stakes through similar approaches and key performance indicators, to the creation of synergies between stakes, moving away from a logic of competition and power struggles. Given the growing importance attributed to other stakes, safety may ultimately benefit more from an articulation perspective than from one based on competition.

Studying road safety policies in Sweden and Norway, Tor-Olav Nævestad and Hanna Wennberg (Volume 2) in their chapter *Lost in sustainability?* offer an excellent example of the current reframing of safety issues. After decades of dedicated public policies, road safety is now included into a wider approach aiming at addressing Sustainable Development Goals such as active mobility, reduction of inequalities, climate and energy use. They discuss a simple, but very relevant question: Is this integration good for road safety, or not? Is it an opportunity ("hitching the wagon" and getting attention and resources) or a threat ("being lost in sustainability" and constrained by a huge, bureaucratic system)? Many other safety issues will face the same dilemma in the short or medium term.

1.4.2.5 Revisit Action: Developing and Maintaining Adaptive Capacity

Several theories and models have outlined strategies for managing unexpected, complex situations that cannot be simply viewed as variations on a nominal scenario. Notable examples include HRO theory and Resilience Engineering. These frameworks offer valuable guidance for developing new foundations for action in a world where uncertainties and complexities are recognized. In these approaches, requirements do exist and serve as references or guidelines; however, it is understood that they are insufficient to support action in certain situations. In such cases, other elements, such as the expertise of individuals (which must extend beyond mere compliance with procedures and processes) and the integration of diverse perspectives, become essential.

In an environment where initiative behaviors and adaptations no longer have to be clandestine, developing and maintaining adaptive capacity might become easier in many respects (e.g., resources, liability aspects).

John Downer (Volume 2), in his chapter *Regimes of fractured objectivity*, points to the limits of the "useful fiction" of objectivity that currently supports the governance of technological risk by states and their regulation bodies. Beyond this epistemic boundary, what does it mean to enter a "regime of fractured objectivity" beyond which expertise loses its power?

Challenges and feasibility conditions

Key elements of HRO theory include redundancies, margins, autonomy, and the empowerment of front-line workers. Similarly, adaptive capacity is a central tenet of the Resilience Engineering approach. These characteristics represent a shift from traditional management models, which typically emphasize financial performance, to a view of organizations that prioritizes safety and adaptability.

Although risk science finds its roots in engineering with seemingly objective calculations and results, Haavik, Antonsen, and Aven (this Volume), in their chapter *Risk, uncertainty and politics: the need for a critical risk science*, reflect on its actual status and needed evolution to meet the challenge of addressing giga-scale phenomena such as climate change or AI for example with impacts on life conditions in societies. The authors highlight why risk science requires to further extend its interactions with other sciences, especially humanities and social sciences. How science develops in these areas of high uncertainty and how it is used are strongly connected to politics. Risk science is no exception, note the authors...

Moreover, this approach to action necessitates a re-evaluation of organizational facades—particularly those directed toward regulators—to ensure that inconsistencies remain manageable. Consequently, this transformation cannot be achieved solely at the organizational level; it must also involve external stakeholders, especially regulators.

Impact on safety

The organizations cited in the development of HRO theory were selected for their exemplary safety performance, despite facing changing demands and operational conditions. Although HRO theory was not intended to be prescriptive, many of its described characteristics are shared by other frameworks for managing complexity and uncertainty. Similarly, fostering adaptive capacity equips an organization with the ability to respond effectively to situations where the environment becomes less predictable or anticipatory, thereby ensuring safety in increasingly uncertain conditions.

The growth in complexity and interconnectedness that is visible in particular in digital infrastructures leads to new safety threats, and to greater consequences when systems break down. New challenges that are inherent to the scale and the multi-layered nature of these systems, which include inter-organizational dependencies whose failure impact has not always been sufficiently considered, require revitalization of the proactive safety capabilities present, as discussed by David Woods (this Volume) in his chapter *Revitalizing safety in crisis.*

1.4.3 Revolution: Divert Discourse and Attention to Innovation Above All Else

This revolution strategy consists in focusing all the attention on innovation, putting at the forefront and above all other consideration the global contributions to society of the new proposed technologies, while disregarding all other aspects, including the potential safety impacts. This strategy involves an overall political, social, and economic environment preventing strong reactions against the damage that could result from accidents.

Such strategy is currently observed as new players are entering high-risk activities with unprecedented radical innovation practices. While anticipation, control, and precaution have historically underpinned the development and operations of high-risk industries and contributed to the emergence of organized hypocrisy, companies such as BYD in China, or Tesla and SpaceX in the USA, appear to rely on fundamentally different premises putting innovation first. "Learning by doing," extensive testing, and the acceptance of failure—including failures that may cause damage, the limits of which are not always clear to external observers—seem to prevail over a priori control facades and traditional safety management approaches. Rapid reaction and adaptation, aimed at integrating lessons learned into subsequent versions, are central to this model in which losses are part of the process. The expectations are so high and the competitive dynamics are so powerful that the requirements of the safety as demonstrated framework appear as an unacceptable hindrance. Whether the expectations are realistic and whether competition is virtuous remain, of course, a question—except for those who are engaged in the "gold rush."

In their chapter, "*AI Gold Rush 2 vs. AI Realism: Overplaying Envisioned Benefits, Underplaying Real Risks*" Rayo, Morey, Grøtan, and Woods (this Volume) refer to the "gold rush" metaphor. They analyze the safety risks posed by the current historic surge in AI, about to include safety-critical activities. They describe the emphasis on overconfident promises, and the underestimation of negative effects. Particularly those destabilizing effects resulting from rushing deployment strategies ("move fast and break things"), bypassing all the precautionary steps traditionally associated with the implementation of technological innovations. They highlight some patterns of instability and vulnerability from both past and current AI Gold Rushes and recommend a "realism" effort to guide the safe and effective deployment of AI-infused technologies. A specific yet powerful illustration of the need to reconcile "safety as promised" and "safety as achieved"!

From a safety perspective, these new players do not seem to feel the need to build explicit safety facades designed to reassure stakeholders. This is made possible, in particular, by the absence or weakness of the regulatory frameworks under which they operate, which eliminates the need to build a facade to reassure regulatory bodies. However, considering the safety performance achieved, especially in the case of autonomous vehicles developed under these conditions, safety is clearly being managed, albeit in ways that differ markedly from traditional approaches, with at least equivalent, if not superior, results. The absence of explicit discourse on safety management removes the very need for organized hypocrisy and constitutes another way of governing safety. As noted by Effron et al. (2018), "an alternative is to avoid words and let deeds speak for themselves" (p. 72), even though, as the authors underline, this deprives organizations of the power of words. Provided there is no unavoidable requirement for public discourse (e.g., legal obligations), safety activities need not be placed at the forefront. Elsbach and Sutton (1992) even suggest that organizations may need to mask controversial activities or divert attention away from them in order to maintain legitimacy. Communication from SpaceX, for instance, places far greater emphasis on the occurrence of a new launch than on the crash of a rocket: Each launch is framed as a success in itself and is given much greater prominence than its safety outcome. A similar pattern can be observed in autonomous vehicle testing, even though some tests have resulted in harm, including to humans.

In the context in the USA at the time of writing, where some regulatory bodies are being dismantled and mandatory facades are therefore no longer required, organized hypocrisy may lose its necessity altogether, potentially leading to very different modes of safety management and governance, as well as to the emergence of new actors.

Challenges and feasibility conditions

While the agile nature of these "opaque" safety management and governance practices, characterized by limited external oversight, may appear better suited to the pace of contemporary change, they raise a number of important questions, including major societal ones, across different time horizons. As Cho et al. (2015) note, "hypocrisy and facade thus allow society the chance to refuse to acknowledge its complicity with the current state of affairs. Society can and perhaps must hold onto higher values than it can live by" (p. 84). Moreover, in contexts where safety regulations do not exist or are no longer in place, questions arise regarding the legal, financial, and societal consequences of accidents. For instance, would the judiciary implicitly assume a fallback safety oversight role in such situations?

Impact on safety

Little information is available on how safety is actually managed in companies following this strategy, and even their safety performance is difficult to assess. The safety performance of autonomous vehicles developed by Waymo in the USA appears to be good, if not superior to that of human-driven vehicles (Waymo/Swiss Re 2025). Further research would therefore be required, either to analyze the actual safety management approaches adopted by these companies in context, or to design and test, at a small scale (for example through regulatory sandboxes), safety management approaches driven by global performance objectives without additional constraining frameworks. Such studies would help to clarify what these approaches might deliver, under what conditions, over which time horizons, and in which types of high-risk domains.

1.5 Conclusion or Starting Point

The perspective of organized hypocrisy adopted in this chapter offers new insights into the inconsistencies currently observed in the management and governance of safety and suggests ways to manage them and to live with them. However, keeping these inconsistencies manageable, while at the same time accommodating ongoing global transformations, may require challenging the current status quo in safety. As illustrated in the previous section, depending on the level of ambition, the scale at which changes are implemented, and the time horizon considered, possible ways forward may not lead to the same outcomes in terms of safety improvement. Although the three strategies discussed earlier are presented as internally consistent and mutually exclusive, empirical reality is far less clear-cut. Measures drawn from different strategic options may be combined, vary from one organization to another, and be adapted to local contexts.

René Amalberti, Siri Wiig, and Charles Vincent (Volume 2) suggest in their chapter *Healthcare in crisis: prospects and directions* that, across industries, valuable lessons can be drawn from the healthcare sector, as it is already confronting the challenge of "living with" intense and persistent pressures. After analyzing these pressures, they identify a range of current adaptations, spanning solutions that remain compatible with the foundations of contemporary industrial safety (e.g., increased reliance on less-skilled workers or the adjustment of quality targets) to more radical approaches (e.g., building on individual autonomy to foster resilience or the large-scale adoption of generative AI technologies). They note that while all these approaches are already implemented to some extent, the most radical ones are neither formally recognized nor evaluated. Studying these practices and strategies may therefore help identify directions for change and improvement that could be relevant beyond the healthcare sector.

Perhaps the key idea in this chapter, beyond applying the "organization of hypocrisy" framework to safety management and governance, is that the current challenges and transformations require action at multiple levels. Though technical and managerial adaptations are of course to be discussed, we need to encompass and reconsider the institutionalized foundations of safety. Whether it is to consolidate them (first strategy), to amend them (second strategy), or to change them drastically (third strategy), living with the new safety landscape requires rethinking the social and political dimensions of safety. If we don't, the crumbling facades will leave us vulnerable to the ongoing upheavals and the risks they entail.

References

E. Abrahamson, P. Baumard, What lies behind organizational facades and how organizational facades lie: an untold story of organizational decision making, in *The Oxford Handbook of Organizational Decision Making* (Oxford University Press, 2008)

M. Baram, C. Bieder, Standardization and risk regulation for high-hazard industries, in *Managing Future Challenges for Safety: Demographic Change, Digitalisation and Complexity in the 2030s.* (Springer International Publishing, Cham, 2022), pp.85–93

C. Bieder, H. Laroche, C. Kamaté, *Public Participation in Governance of Industrial Safety Risks: An Uneasy Journey* (Springer Nature, 2025), 98 p

M. Bourrier, C. Bieder, Trapping safety into rules: an introduction, in *Trapping Safety into Rules* (CRC Press, 2017), pp. 1–9

N. Brunsson, *The Organization of Hypocrisy: Talk, Decisions and Actions in Organizations* (Wiley, Chichester, 1989)

N. Brunsson, Ideas and actions: justification and hypocrisy as alternatives to control. Acc. Organ. Soc. **18**(6), 489–506 (1993)

N. Brunsson, *Mechanisms of Hope: Maintaining the Dream of Rational Organization* (Copenhagen Business School Press, Copenhagen, 2006), p.239

N. Brunsson, *The Consequences of Decision-Making* (Oxford University Press, New York, 2007), p.174

C. H. Cho, M. Laine, R. W. Roberts, M. Rodrigue, Organized hypocrisy, organizational facades, and sustainability reporting. Acc. Organ. Soc. **40**, 78–94 (2015)

M. Crossan, M. P. E. Cunha, D. Vera, J. Cunha, Time and organizational improvisation. Acad. Manage. Rev. **30**(1), 129–145 (2005)

EAIB (Ethiopian Airplane Accident Investigation Bureau), Aircraft Accident Investigation Report B737- MAX 8, ET-AVJ (December 2022). https://bea.aero/fileadmin/user_upload/ET_302__B737-8MAX_ACCIDENT_FINAL_REPORT.pdf#page=198&zoom=100,62,122

D. A. Effron, K. O'Connor, H. Leroy, B. J. Lucas, From inconsistency to hypocrisy: when does "saying one thing but doing another" invite condemnation? Res. Organ. Behav. **38**, 61–75 (2018)

K. D. Elsbach, R. I. Sutton, Acquiring organizational legitimacy through illegitimate actions: A marriage of institutional and impression management theories. Acad. Manage. J. **35**(4), 699–738 (1992)

C. M. Frantz, F.S. Mayer, The emergency of climate change: why are we failing to take action? Anal. Soc. Issues Public Policy **9**(1), 205–222 (2009)

I. Fucks, Y. Dien, 'No rule, no use'? The effects of over-proceduralization, in *Trapping Safety into Rules* (CRC Press, 2017), pp. 27–39

C. Gendron, Penser l'acceptabilité sociale: au-delà de l'intérêt, les valeurs. Revue internationale de communication sociale et publique, (11), 117–129 (2014)

M. Griffin, Competencies and skills for navigating uncertain and interdependent systems of the future, in The Shaken Foundations of Safety: Overcoming Human Resources and Skills Shortages (Springer, Cham) (2026).

Grignon et al., *Criminalization of Errors and Deviations from Procedure in Post-accident Inquiries.* Number 2025-01 of the Cahiers de la Sécurité Industrielle (Foundation for an Industrial Safety Culture, Toulouse, France, 2025). https://doi.org/10.57071/rgr212. Available from FonCSI.org/en

A. Hale, D. Borys, Working to rule, or working safely, in *Trapping Safety into Rules* (CRC Press, 2017), pp. 43–68

S. T. Hansen, S. Antonsen, Taking connectedness seriously. A research agenda for holistic safety and security risk governance. Saf. Sci. **173**, 106436 (2024)
E. Hollnagel, D.D. Woods, N. Leveson (eds.), *Resilience Engineering: Concepts and Precepts* (Ashgate Publishing, Ltd., 2006)
E. Hollnagel, *Safety-I and Safety-II: The Past and Future of Safety Management* (CRC Press, 2018)
J. C. Le Coze, B. Journé, *Compliance and Initiative in the Production of Safety* (Springer Nature, 2024a)
J. C. Le Coze, B. Journé, *The Regulator–Regulatee Relationship in High-Hazard Industry Sectors: New Actors and New Viewpoints in a Conservative Landscape* (Springer Nature, 2024b), p. 114
T. R. La Porte, High reliability organizations: unlikely, demanding and at risk. J. Contingencies Crisis Manag. **4**(2), 60–71 (1996)
H. W. Lim, H. Lingard, R. P. Zhang, P. Pirzadeh, *Decluttering Safety: A Multistakeholder Approach to Enhancing Safety Management in Construction* (RMIT University, 2025)
V. Milch, K. Laumann, Interorganizational complexity and organizational accident risk: a literature review. Saf. Sci. **82**, 9–17 (2016)
P. C. Nystrom, W. H. Starbuck, Organizational façades, in *Academy of Management Proceedings* (Academy of Management, Briarcliff Manor, 1984), Vol. 1984, No. 1, pp. 182–185
D. Penklis, *Organized Hypocrisy* (Troubador Publishing Ltd., 2024)
M. Power, *The Audit Society: Rituals of Verification* (Oxford University Press, 1997)
M. Snyder Caron, A. Gupta, The social contract for AI. arXiv e-prints, arXiv-2006 (2020)
K. Størkersen, T. Thorvaldsen, T. Kongsvik, S. Dekker, How deregulation can become overregulation: an empirical study into the growth of internal bureaucracy when governments take a step back. Saf. Sci. **128**, 104772 (2020)
S. Tillement, C. Kamaté, H. Laroche, C. Bieder, *The Shaken Foundations of Safety: Overcoming Human Resources and Skills Shortages* (Springer, Cham, 2026)

Part II
New Frames for a Changing World

Chapter 2
Revitalizing Safety in Crisis

David D. Woods

Abstract This paper highlights cycles of crises in safety: crisis, shock, possible revitalization, destabilization, recurring crisis. Once the pattern is visible, the cycle can be "hacked" via techniques to create foresight. As forces of change accelerate, the stream of incidents and accidents that follow reveal the loss of vitality of safety-related capabilities as in two exemplar events. Acting ahead of breakdowns forestalls impending crises and dampens resultant shocks in terms of harm and loss of effectiveness. New techniques can convert easy-to-discount events that signal ongoing *devitalization* into demand signals that reveal *paths for revitalization*. The *flying reserve* technique is one strategy that organizations can implement immediately.

Keywords Safety in crisis · Safety of complex systems · Adaptive capacity · Resilience engineering · Flying reserve · Proactive safety

2.1 Crisis

A fundamental obligation in safety is to "create foresight about the changing shape of risk before harm occurs" (Woods 2005; Woods et al. 2025). Being proactive is anticipating how vulnerabilities change and new threats emerge, recognizing how past safety initiatives lose vitality, and empowering people to revise systems before harm occurs. Given the dramatic changes afoot in societies and technologies, our ability to carry out complex and risky activities safely and effectively is in crisis undermining our ability to meet this basic obligation (Bieder et al. 2022).

Crises in the pursuit of safety have occurred before as other priorities overwhelm the ability to be proactive. I first encountered crises in doing things safely 47 years ago in the shock that followed the 1979 Three Mile Island nuclear accident. Prior to the accident, authorities and engineers were overconfident that risks were well-understood and the mechanisms developed to counter those risks were effective—nuclear power had reached "safe" status. The shock wave from this and later failures

D. D. Woods (✉)
Ohio State University, Columbus, OH, USA
e-mail: woods.2@osu.edu

C. Bieder et al. (eds.), *Living with the New Safety Landscape I*,
SpringerBriefs in Safety Management, https://doi.org/10.1007/978-3-032-29336-7_2

in aviation, space, transportation, and health care triggered many to re-examine and develop new models to revitalize creating safety in an increasingly complex world.

Clausewitz diagnosed an early example of a cycle of crisis, shock, and revitalization this way: "A vain, immoderate faith in these institutions made it possible to overlook the fact that their vitality was gone. The machine could still be heard clattering along, so no one asked if it was still doing its job." (Clausewitz 1992, p. 32). Although he was referring to the shock collapse of Prussian society to Napoleon in 1806, the patterns he experienced and described echo today: Overconfidence in past advances blocks recognition of a changing world and blinds organizations to clear evidence past mechanisms have become stale/ineffective. Resilience Engineering recognized the same patterns in safety of complex systems in modern times: Systems become increasingly brittle, so that as new challenges arise, they fracture. Importantly, Clausewitz then and Resilience Engineering now have charted paths to revitalization, paths which build the adaptive capacity to flourish as change, growth, and challenge continue (Woods, 2024, 2026).

2.2 Failures of Foresight in Proactive Safety

The ongoing loss of vitality in safety is illustrated in the January 29, 2025, midair collision in Washington DC. The NTSB March 11, 2025, preliminary report (https://www.ntsb.gov/investigations/Documents/DCA25MA108Prelim.pdf) highlighted that near misses were a common occurrence for years between commercial traffic landing at DCA airport and government/military helicopter traffic along the Potomac River (e.g., collision avoidance warnings occurred regularly, some requiring commercial aircraft to fly go arounds to avoid helicopters).

How could this longstanding vulnerability have been missed when aviation had developed independent incident reporting systems to recognize vulnerabilities before accidents harmed people, such as the Aviation Safety Reporting System (ASRS). The ASRS itself is a proactive safety innovation that was the product of revitalization in the face of a shocking accident in the mid-1970s despite organizational resistance to change.

How did aviation incident reporting systems fail to alert the aviation industry before such a consequential accident occurred? Prior to the 2025 midair collision, collection mechanisms "clattered along" documenting aircraft/helicopter close calls. Yet no role or organization integrated the reports; no one synthesized the information to highlight a pressing vulnerability; and no one led action to reduce the danger by reconciling the conflicting priorities of different authorities to revise how traffic streams flowed near the airport. Multiple roles and parties stayed fragmented within silos, leading to this breakdown in proactive safety.

This accident also was a remarkable due to the failure to apply lessons gained from many past accidents (Woods 2005) and highlighted in multiple analyses by leading safety advocates James Reason, Diane Vaughan, and Richard Feynman among others (Herald of Free Enterprise ferry, Challenger and Columbia space shuttles, BP Texas

City refinery and Deep Water Horizon oil platform explosions). In those accidents, as in this one, the clear evidence of trouble ahead was discounted and rationalized away even as the system operated on the brittle edge of failure.

The need for organizations to adapt to changing vulnerabilities as capabilities grow is also visible in the October 20, 2025, AWS (Amazon) 15-h outage of vital digital services[1] which underpin a wide swath of critical activities in multiple sectors. This is one exemplar of new forms of risk and harm—Wide Swath Outages of Critical Digital Services.

The outage arose from the extensive interdependencies that span multiple scales—connecting technical aspects of the software layered "stack" with the layers of human activities all the way up to societal scales. This tangle arises from growth of valuable capabilities, in this case cloud computing run by an organization with enormous resources. As a result, the outage impacted hundreds of services and activities of varying criticality across human stakeholders. The extensive tangle of interdependencies undermined the function of critical digital infrastructures themselves leading to failures across different industrial, financial, educational, and transportation sectors. The outage affected many other critical digital services. These secondary effects then fed back on the triggering outage in several ways adding anomalies, overload, bottlenecks, uncertainty to complicate assessment and response. The event also disrupted AI LLM services (Anthropic/Claude; Perplexity) both directly and indirectly since these services depend on other software infrastructure affected by the AWS outage. Activities which utilized these services were impaired.

Software operations underpin virtually all valued activities in all sectors. Events like this one illustrate how the field of software operations handles a continuing stream of incidents that threaten outages before harm occurs despite scale, criticality, and autonomy of the underlying processes (Allspaw 2022). Inevitably given the complexity, some challenge events evolve to outages and harm before interventions return valued services. Organizations with enormous resources are not immune.

2.3 The Changing Shape of Risk and Harm

These events highlight the challenge for systems safety in a changing world—can we keep pace with the changing shape of risk and harm (Woods, 2020)? The two breakdowns exemplify ongoing shocks that break or exceed the capacity of current safety initiatives. Two of many complexity effects are:

---

[1] It is easy to get lost in the unfamiliar terrain of software operations and the technical nature of the processes being operated (e.g., what's a DNS?). Software operations and the processes managed exhibit deep general patterns about resilient performance and safety in complex highly autonomous, systems that function at scale.

https://incident.io/blog/service-disruption-october-20th-2025.

https://www.thevoid.community/newsletter.

https://www.planetearthandbeyond.co/p/amazon-just-proved-ai-isnt-the-answer.

https://surfingcomplexity.blog/2025/10/25/quick-thoughts-on-the-recent-aws-outage/.

Scale Effects Magnify Harm. Organizational and technological efforts to grow capabilities expand the network of interdependencies over multiple roles, layers, organizations. Capabilities' growth provides attractive benefits. It also expands how trouble arises, what is needed to respond effectively, and the consequences of breakdowns. Past methods by themselves cannot cope with the new scale of systems, failure, and harm that results.[2]

Coordination Is Basic for Safe and Effective Performance. As Hochstein highlights, all modern systems are distributed and layered, and thus all challenges are ones of coordination and *synchronization* (https://bsky.app/profile/norootcause.surfingcomplexity.com/post/3l4hdlt4zpk2m).

The technology and software layered "stack" of technologists have become intertwined with multiple layers of human activities all the way up to societal scales (Haavik 2020). All problems now are challenges of synchronizing activities over roles and layers to keep pace with change. The scale of these activities has expanded organizational and jurisdictional lines drawn by human interests. The penalties for fragmentation across roles and layers are illustrated by both exemplar failures.

While synchronization dominates, most work clatters along stuck in a narrower focus on single human roles babysitting/supervising/interacting with a software/machine agent (though underneath all are distributed systems). Much less attention flows to designing for synchronizing joint activity distributed over a space of interdependent roles, multiple human and machine/software actors, and situated in a changing world where multiple threads of interleaved activities are intertwined over time (Woods et al. 2025).

Systems operating at new scales fail differently. Yet tired ideas persist: the belief that failures arise from a singular or root cause misrepresents how complex layered infrastructures function and malfunction. Triggering events often are not a failure at all. Failures of a component or subsystem normally do not lead to outages of the valued services. The growth and criticality of the layered infrastructure is only possible if the system is robust in the face of subsystem degradations. Incidents that threaten outages are emergent effects of the complexities that accompany growth, as illustrated in the AWS outage. Pragmatically, a new profession of software operations (Continuous Integration/Continuous Deployment) has grown up in the last 15 years to manage this critical digital infrastructure which utilizes techniques and expands findings of Resilience Engineering (Cook and Long 2021).

Previous methods to build safety have been superseded by bigger, more powerful dynamics operating over much broader scopes and producing negative effects several *orders of magnitude larger* than the most optimistic positive impact of past safety initiatives (Woods et al. 2025). The scale effects of growth and accompanying

[2] Complexity science uses scale like biology in the sense that growth produces emergent systems that operate at a new scale—new processes, relationships, and laws (e.g., sub-cellular processes operate at a different scale than organ systems). Others use the label scale to refer to increasing performance on a dimension by increasing the magnitude/mass of a resource (e.g., AIers claim increasing the mass of data and computation will keep increasing intelligent capabilities).

complexity penalties demand a new unifying synthesis across all the different disciplinary, technical, human, and engineering perspectives, each of which brings a partial, valuable lens. Fragmentation across these lines of inquiry is a major barrier. Simply juxtaposing several perspectives is insufficient since the tangled interdependencies defy complete mapping. Hidden lines of dependency, circular dependencies, and even strange loops arise and can only be seen when layered systems operate under stressful real-world loads—the AWS outage. New scientific fundamentals offer hope to capture the emergent dynamic properties of growth, complexification, and adaptation (Woods 2026).

2.4 Revitalization

Organizations can build the necessary adaptive capacities. Revitalization strategies innovated in Clausewitz's time are valuable whenever multiple destabilizing forces generate turbulence. Turbulence arises today from the interplay between technological change (Rayo et al., 2026), societal fragmentation (Carney https://globalnews.ca/news/11620877/carney-davos-wef-speech-transcript/), and extreme environmental stresses.

Revitalization begins with honest examination of the gap between stale beliefs and reality as revealed in the exemplar breakdowns above. Assumptions of almost linear systems, human-caused failures, and that more automation always reduces risks are untenable as complexity effects expand as a result of the same forces that produce growth (Woods 2026; https://resiliencefoundations.github.io/video-4-the-science-and-pragmatics-of-re-through-the-lens-of-complexification.html). Vulnerabilities to failure have changed, and new threats have emerged.

Revitalization requires building more powerful and more adaptive mechanisms for coordination and mutual assistance across organizational boundaries to anticipate and build responsiveness to acute and chronic stressors. In one of the exemplars, breakdowns across organizational boundaries led to the failure of proactive safety. In the other, the breakdown in critical services triggered a spiral of breakdowns that degraded a wide swath of other critical services and valued activities for stakeholders. Flexible multi-lateral relationships are needed to be highly responsive, yet still thorough, when challenges threaten overload and the risk of harm rises. The ability to re-prioritize when disruptions intensify goal conflicts changes which normal goals need to be relaxed/sacrificed relative to others. The ability to re-configure relationships across roles and layers allows synchronizes of activities to keep pace with events. Both are core strategies of Resilience Engineering (https://resiliencefoundations.github.io/video-11-beyond-surge-capacity.html).

The paradox of times of turbulence is: *extra investments in adaptive capacity are most needed when they are least affordable*, trapped between pressure to be responsive to change and pressures for reducing costs. Escaping the trap requires re-architecting the system to be skilled at the art of re-balancing goals/reconfiguring activities.

Organizations can build flexible multi-lateral relationships to enhance these forms of adaptive capacity. Many strategies have been documented and explored. One technique for resilient performance at scale can be achieved via a maximally flexible *flying reserve* of human experience/technical expertise. "Flying" means able to move promptly to meet specific challenges as they arise, wherever they arise. "Reserve" means ready-at-hand potential for action at decisive times/junctures. It is a synchronizing unit, drawn flexibly from many standard units, that cuts across roles and layers. This adds extensibility to handle surprise and overcome vulnerabilities from fragmentation that are evident in the two examples.

Recognizing when and where to deploy the flying reserve is critical to seize opportunities to improve *both* effectiveness and safety. It must have sufficient autonomy, discretion, and independence from orthodoxies of organizational practice to be maximally flexible. It is staffed and assembled opportunistically, applying a tailored mix of experience fit to the issues that arise. It cuts across multiple silos, bringing know-how from across diverse organizational activities into a function that boosts coordination, which is increasingly essential as interdependencies dominate performance of modern systems.

Two examples reveal the flying reserve's feasibility and criticality. Following the Columbia Space Shuttle accident, NASA created the NASA Engineering and Safety Center (NESC). It has functioned to improve *safety and effectiveness* for over 20 years. The NESC assembled a customized team, including external specialists in Resilience Engineering, to follow up a near-miss incident on a spacewalk. Their work uncovered a myriad of problems for spacewalks, mission control, and space station operations that generalized to many other NASA activities, all missed by a standard RCA (Holbrook et al. 2025). The findings addressed both safety and effectiveness because these goals are so tightly intertwined in NASA operations. In contrast, Boeing demonstrated the consequences of compartmentalizing technical expertise in silos in twin 737 Max accidents in 2018/19. By prioritizing schedule and financial risk over thorough engineering/testing, the engineering modifications contained flaws leading to 346 deaths.

2.5 Summary

Revitalization of past practices is difficult when the world is changing dramatically. Yet organizations need to adapt to keep pace with the changing shape of risk and harm. Systems operating at new scales fail differently with orders of magnitude larger consequences. NASA demonstrates the feasibility of the flying reserve strategy, exemplifying how organizations can enhance the ability to act decisively as they recognize emerging risks to safety and effectiveness. The flying reserve acts as a flexible synchronizing unit to overcome vulnerabilities from fragmentation—a major risk seen in failures of proactive safety.

References

J. Allspaw, Critical digital services, in *Managing Future Challenges for Safety: Demographic Change, Digitalisation and Complexity in the 2030s*, ed. by H. Laroche, C. Bieder, J. Villena-López, (Springer International Publishing, 2022), pp. 33–40. https://doi.crg/10.1007/978-3-031-07805-7_4

C. Bieder, J. Villena-López, Times are changing and so is safety, in *Managing Future Challenges for Safety: Demographic Change, Digitalisation and Complexity in the 2030s*, ed. by H. Laroche, C. Bieder, J. Villena-López, (2022), pp. 1–12. https://link.springer.com/chapter/10.1007/978-3-031-07805-7_1. Or both Bieder et al. this volume

R.I. Cook, B.A. Long, Building and revising adaptive capacity sharing for technical incident response: a case of resilience engineering. Appl. Ergon. **90**(2) (2021). https://doi.org/10.1016/j.apergo.2020.103240

T.K. Haavik, Societal resilience: clarifying the concept and upscaling the scope. Saf. Sci. **132** (2020). https://doi.org/10.1016/j.ssci.2020.104964

J.B. Holbrook et al., *Best Practices for Organizational Resilience in the International Space Station (ISS) Program*. NASA/TM−20250005960/NESC-RP-14-00990 (2025). https://ntrs.nasa.gov/api/citations/20250005960/downloads/6-9-2520250005960.pdf

M.F. Rayo, D.A. Morey, T. O. Grøtan D.D. Woods, AI Gold Rush 2 vs. AI Realism: Overplaying Envisioned Benefits, Underplaying Real Risks, in Living with the new safety landscape: Re-thinking safety management in the context of cascading uncertainties, ed. by C. Bieder, E. Marsden, H. La Roche, (Springer Open & the Foundation for Industrial Safety Culture, Springer Briefs in Safety Management, Toulouse France, 2026).

C. von Clausewitz, in *Historical and Political Writings*, ed. by P. Paret, D. Moran, (Princeton University Press, Princeton, NJ, 1992), p. 32

D.D. Woods, Creating foresight: Lessons for resilience from Columbia, in *Organization at the Limit: NASA and the Columbia Disaster*, ed. by W.H. Starbuck, M. Farjoun, (Blackwell, Malden, MA, 2005), pp. 289–308

D.D. Woods, The strategic agility gap: How organizations are slow and stale to adapt in a turbulent world, in *Human and Organizational Factors: Practices and Strategies for a Changing World*, ed. by B. Journé, H. Laroche, C. Bieder, C. Gilbert, (Springer Open & the Foundation for Industrial Safety Culture, Springer Briefs in Safety Management, Toulouse, France, 2020), pp. 95–104. https://doi.org/10.1007/978-3-030-25639-5

D.D. Woods, The command-adapt paradox: Guided adaptability to cope with complexity, in *Compliance and Initiative in the Production of Safety: A Systems Perspective on Managing Tensions and Building Complementarity*, ed. by J.-C. Le Coze, B. Journé, (Springer Open & the Foundation for Industrial Safety Culture, Springer Briefs in Safety Management, Toulouse France, 2024) https://link.springer.com/chapter/10.1007/978-3-031-45055-6_8

D.D. Woods, M.F. Rayo, S. Pruchnicki, The Silicon Valley way: move fast and break... aviation safety? Bull. Atomic Sci. **29**, 2025 (2025) https://thebulletin.org/2025/05/the-silicon-valley-way-move-fast-and-breakaviationsafety/

D.D. Woods, Resilience Engineering: A Radical Reframing in Three Parts. in *Resilience Engineering: Progress, Challenges and Opportunities*, ed. by A. Adriaensen, R. Patriarca, G. Praetorius, (Springer Briefs in Safety Management, 2026) in press

Chapter 3
Risk, Uncertainty, and Politics: The Need for a Critical Risk Science

Torgeir Kolstø Haavik, Stian Antonsen, and Terje Aven

Abstract The handling of societal risks takes place at the boundaries between science and politics, and between several scientific domains. We distinguish between "research *in* risk science" taking place at the "core" of risk science—and "research *on* risk science" taking place at the marginal zone of risk science, in close communication and negotiation with other fields. Facing giga-scale and transformative challenges like climate change and the proliferation of AI, a critical risk science needs to be reflexive. In addition to solving problems formulated by others, it must also actively and critically engage with problem formulations, e.g., challenges arising when risk problems transcend the boundaries of disciplines, levels of analysis and existing governance structures. Ensuring critical dimensions of risk science to meet such challenges, requires an exploration of the intersection to other sciences, particularly the humanities and social sciences, to strengthen what has been labeled *research on* risk science.

Keywords Research *on* risk science · Research *in* risk science · Uncertainty · Societal risk

3.1 Introduction

What are the research challenges for safety science in an increasingly complex world, where unforeseeable consequences of rapid technological, social, and political developments may unfold at an increasingly large scale?

T. K. Haavik (✉)
NTNU Social Research, Trondheim, Norway
e-mail: torgeir.haavik@samforsk.no

S. Antonsen
Statnett, Oslo, Norway

T. Aven
University of Stavanger, Stavanger, Norway

C. Bieder et al. (eds.), *Living with the New Safety Landscape I*,
SpringerBriefs in Safety Management, https://doi.org/10.1007/978-3-032-29336-7_3

We want to start with a premise. The increasing focus on security the last decade or so has challenged safety science to expand its traditional scope of theorizing, preventing and coping with unintended consequences to embrace also intended consequences. Further, the central subject of safety science has been consequences that unfold within the scale of organizations or sectors. This scale has been expanded following the increasing focus on *societal security*, which in the Norwegian safety literature has been labeled *societal safety*.[1] With this two-dimensional expansion of scope for safety science, the notions or risk and uncertainty have emerged as increasingly central common denominators for both safety and security research across societal scales. Although these terms are by no means new to safety science, a need has emerged to carefully relate or align "traditional" safety science with other scientific fields and scopes to keep both relevance and integrity when new scientific needs emerge in society. With increasing complexity of systems and threats, many challenges cannot be solved by a "pure" safety focus alone, but we also contend that it may cause more harm than good if a number of threats are addressed as cases of security alone, without respecting the implications with respect to safety.

Against this backdrop, we acknowledge that meeting the challenges for safety science requires meeting the challenges for a larger, multi-disciplinary risk science, under which both safety research and security research, at different scales, can be meaningfully sorted and also find broader application. Therefore, we believe that the question "what are the research challenges for safety science in an increasingly complex world," is best answered if reformulated into *what are the research challenges for risk science in an increasingly complex world?*

This chapter is devoted to addressing this question. Black boxes located in the terra incognita at the marginal zone of risk research will be inspected, to discuss how knowledge fundaments of mutually agreed-upon concepts and ideas from risk science fall short of explanatory potential when confronted with unprecedented risks and high uncertainty.

Unlike the more mature areas of risk research that has been developed, contested and (more or less) settled through years, the marginal zone of the field represents an area where new risk phenomena are addressed and new scientific developments are taking place, hence scientific uncertainty is high. Over time, there is also a movement from a scientific field's periphery, toward its established core. As such, research topics, theoretical considerations and cases studied at the margins of the field contains high potential for scientific innovation as the field's current fringes may be its future core.

The lack of knowledge thus invoke both ontological and epistemological considerations, with all the sources of uncertainty that science in the making involves, and that Latour (2005) has accounted so thoroughly for, in play.

---

[1] Just to complicate, in Norway we use the term "societal safety," although at the societal level, Norwegian research literature does not distinguish as consequently between safety and security as the research literature at the organisational level does.

In the following, we shall discuss uncertainties associated with interdisciplinary development of new concepts and knowledge of risk science in a context of societal change and large-scale risks.

3.2 Research *in* Risk Science and Research *on* Risk Science

Many of the core ideas or concepts in risk science were developed at the intersection to other sciences. Think of probability, black swans, sense-making, requisite variety, or safety culture; these are but a few examples of concepts applied in risk science today, but whose origins can be traced back to psychology, anthropology, cybernetics, and statistics, among others. A result of the continuous boundary work taking place at the borders between scientific fields or research traditions is sometimes that concepts travel across borders and become internalized or domesticated as a core concept in a field.

In a recent paper, Antonsen et al. (2025) distinguish between "research *in* risk science" taking place at the "core" of risk science, in comfortable alignment with established discourses and scientific journals—and "research *on* risk science" taking place at the marginal zone of risk science, in close communication with other fields, specifically covering reflexive research that enables translation and circulation of knowledge between the inside and the outside of risk science. The core of risk science is here understood as concepts, principles, frameworks, approaches, methods, theories and models for assessing, characterizing, communicating and handling risk, in general and for specific applications (Aven 2018; Hansson and Aven 2014).

Research on risk science represent examples of boundary work, and is an interesting case enabling studies of research fields *in the making*, for example—of particular interest for this audience—risk science in the making.

We argue that the field of risk research should be understood in a context of other fields it is indebted to. Often the borders between the fields are fuzzy, and they tend to be fruitful with respect to new scientific concepts coming into being, before they (sometimes) travel toward field cores to become part of the "standard" repertoire of that field.

This way of thinking about field development finds resonance in the acknowledgment that—as well as it can explain mechanisms through which—"crises" may represent opportunities for scientific development.

Crises come in different forms. Some crises are constituted by a situation demanding more resources and efforts than are available to, e.g., an organization or a society. Other crises are of a more foundational type, challenging both existing ontologies and epistemologies by being unprecedented or unrecognizable phenomena. In the former type, uncertainty is low. In the latter, uncertainty can be extremely high. We may call this latter type of crises "foundational crises."

Uncertainty is higher in the marginal zone of research fields than in their cores. Put another way, there is a lack of knowledge on those phenomena that have not (yet) found "residence" in, and have not been conceptualized by, a scientific field.

One concrete example of this is the problem complex of security of offshore infrastructure and energy supply when (hybrid) warfare emerges and produces a new, mixed context spanning from organizational safety to geopolitical security (Hansen and Antonsen 2024), This novel risk traverses companies, governance and industrial sectors in a manner that begs for new developments of scientific comprehension and conceptualization.

Potentially creeping crises, including new, groundbreaking technologies whose consequences we cannot yet oversee, tend to be situated in a scientific landscape of high uncertainty. This goes for climate change, it goes for national security when long-lasting political institutions crumble, and it goes for AI.

Climate change, for example, might neither be understood nor solved without knowledge and solutions being harmonized across fields of, e.g., renewable energy, international policy development and consumption psychology. For each of these fields, the challenge is new in the sense that it exists in a particular context of strong interdependence with the other fields' challenges. For a risk researcher, thus, climate risk cannot be reasonably dealt with in isolation from these other fields.

On Risk, Uncertainty, and Politics

Risk science applications have long and strong positivist dispositions, in the sense that the field was for a long time associated with efforts to objectively *predict* future risks, in order to meet them with insurance schemes or showing that risks were acceptable in relation to the benefits produced by economic activity. In addition, quantification of risk has played a crucial role in planning and decision-making, in order to achieve a desired outcome with as little risk as possible.

However, there has also been strong counterforces to treatment of risk as merely "objective" input to decision-making. Leading risk scholars like Paul Slovic, Ortwin Renn, George Apostolakis, and Elisabeth Paté-Cornell all rejected such positivist dispositions, stressing the need for broader perspectives to understand, assess, communicate and handle risk (Aven 2025a). Knowledge on risk perception, risk assessment, risk communication, and risk management (governance) has developed, providing a foundation of risk analysis and risk science which clarifies the basis of the field, making a sharp distinction between the risk science—the practice of developing the most justified knowledge of the field and discipline—and the actual risk analysis (perceptions, assessments, communication, handling risk) in practice. Ulrich Beck's risk society (Beck 1992) is an example of the latter.

The sociologist Ulrich Beck worked with understanding the role of risk in modernity. In his sociology of modernity, Beck (1992) analyzed risk more as a phenomenon that is at the same time both a product of late modern societies, and allowing the introduction of hazards and insecurities to society. The risk framing comes with an implicit promise of predictability and control. In that perspective, risk is seen as a central organizing principle for social life, to such a degree that we may talk about a risk culture, a risk society. Risk is treated not as external and unpredictable, but as "internal" (man-made) and manufactured, intrinsically linked to technological advancement. In the reflexive modernity, thus, any discussion of risk reflects as much culture, institutions and perceptions as they reflect how risk is framed by risk experts (Douglas and Wildavsky 1983; Glasner 2013).

The ideas of Beck on risk society illustrate that risks that are of societal or global scale are not about something "inert" that can be objectively measured or described. In Beck's conceptualization of risk, there is no clear distinction between the concept of risk and the management of risk. This is strongly related to his emphasis on risk analysis as coming with a promise of the risks described being something that can be managed. Although Beck goes far in equaling technical risk analysis with an ideological strategy to "sell" risk acceptance to the public (Renn 2008), there is an inherent duality in technical risk analysis: At the same time as it serves to reduce risk, it also serves to legitimize decisions to accept and take the (residual) risks in question. In this way, there is a link between the ideally "clean" world of risk science and risk description, and the messier world of public discourses of risk and risk governance. This link indicates that the field of risk is inherently political insofar as it also serves a function in policy development in society in connection with large-scale hazards with considerable uncertainty.

Importantly, the link between the two domains is bi-directional. Not only does the scientific domain influence public discourses and governance of risk, but also the other way around in terms of influencing the framing of what is seen as dangerous and safe, which topics deserve research funding, and whose voices receive the most attention in both these matters. This is why also the assessment of risk is "necessarily a social and political exercise, even when the methods employed are the seemingly technical routines of quantitative risk assessments" (Jasanoff 1999, p. 150).

Clearly, the actual use of risk analysis and risk science in practice—is political, as politicians make decisions involving risk-giving weight to all relevant concerns and values. To what extent risk science with its foundation in concepts, methods, theories and models is political, is however a different question. It is related to the more general question about science being political, which has been thoroughly discussed in the literature over the years (e.g., Brown 2015; Büntgen 2024).

Scientific processes have subjective (inter-subjective) elements and are dependent on values. However, the scientific processes follow some common standards leading to a fundamental difference between science and politics: Science provides knowledge about nuclear power plants and associated risks, whereas politics also include judgments on how you like or dislike these risks which extend beyond the scientific domain.

This description of the messy relation between science and politics does not preclude the normative stance that the boundaries of the two domains should be defended. For instance, Renn (2008, p. 11) argues that "[t]he fact that evidence is never value-free and that values are never void of assumptions about evidence does not compromise the need for a functional distinction between the two." In this perspective, the boundaries should be defended, since neither the politization of science nor the technocratization of politics is desirable within democracies. The analysis and management of societal risks take place at the boundaries not only between science and politics, but also between several scientific domains. In this view, (ontological and epistemic) uncertainty with respect to understanding or managing risks increases with the distance a risk phenomenon is from the core of a scientific discipline and the discipline's familiarity with the risk phenomenon. Uncertainty is truly an ambivalent entity in connection with risk governance and decision-making. A lack of knowledge tends to invite extensive knowledge acquisition. This is a process often postponing decisions to deviate from or halt the current direction of, e.g., technology development or societal development trends, and by that continuing development paths that produce dependencies that are increasingly difficult to reverse. Human-induced climate change and AI development can serve as examples.

This can be one contemporary reading of Ulrich Beck, and the dynamics of the risk society. Acknowledging this phenomenon as a risk of second order,[2] risk science can be a useful instrument in questions of technology development and societal developments, although they involve questions that are inherently political.

The division of labor between science and politics, it can be argued, is possible only in an "ideal world," just like many ideal "laws" of physics—e.g., Newton's gravitational law and Einstein's law on special relativity that are "valid" only in empty space. However, these laws are also useful in practice despite their limitations, and the same logic we will argue applies for science in general. Risk science knowledge cannot be seen isolated from the context it is produced, yet it can provide insights that are not linked to specific political stances.

When it comes to *science in action*, scholars of Science and Technology Studies have demonstrated thoroughly that the use of science *is* inherently political, a viewpoint encapsulated in Latour's famous quote "Science is politics by other means" (Latour 1988). The use of risk science (the "risk world") is no exception, being occupied with the investigation of "the consequences of activities, with respect to *something that humans value* (…)" (Aven and Renn 2009, our italics). These things that humans value, of course, refer to the same things that are incorporated in the famous definition of politics, that "Politics is the authoritative allocation of values in society" (Easton 1953). Science and risk science in particular are however not about allocation of values, but knowledge supporting this allocation. This knowledge and its related knowledge generation are to varying degree influencing and are influenced by politics.

The slogan of NTNU, for example, is "Knowledge for a better world," and no wonder that many of the scientific related battles being at their deepest about what a better world would look like. With changing political battles, the scientific related battles changes. To scientifically describe the edge of the Arctic ice, meaning the border between the open sea and the Arctic Sea ice, political issues were involved restricting the oil and gas industry's expansion into Arctic areas. The scientific work was influenced by politics. It can be argued that the issue of determining the edge of the Arctic ice is not mainly about science but values and hence politics.

---

[2] By second order we mean that another and more serious risk may develop in the wake of more immediate risks being "resolved." A hypothetical example could be the ethical "primary" risks of AI. These may be resolved at the "micro scale," and resource- and energy-intensive AI may thus continue to be irreversibly integrated in almost all aspects of life, with the large-scale second-order risks of nature loss and climate change.

A phenomenon like climate change and the associated climate risks, can be addressed from a risk science perspective, for example on how to characterize and communicate climate risk, on how to understand the precautionary principle in relation this type of risk, and studies on how current risk policies work in practice. Actual risk handling of climate change risk issues extends however far beyond this type of work. In order to fully understand the space of maneuver and the requirements for prevention, mitigation, and adaptation, there is a need to include perspectives and methods from other fields like political science, sociology, anthropology, and psychology. Such work can be labeled applied risk analysis (science) when risk is a core issue (Aven 2018), but it can also be referred to as climate change research.

Analysts and researchers working on climate change may identify themselves as climate scientists (analysts), but at the same time recognize the existence of an overriding science on risk topics—risk science—covering the most justified knowledge and practices in relation to risk understanding, assessment, communication and handling (Aven 2025b).

Climate risk can be situated as an emerging topic in a marginal zone of and between several sciences, where a number of central research issues are yet unresolved: what types of groups and institutions exist, emerge or are disappearing that may make up relevant units of investigation or (local interest organizations, BigTech, IMF, IPCC, UNFCCC, WWF); how far along the networks of actors should impact and agency be traced (local community, EU, UN); who should be considered legitimate actors and stakeholders (e.g., individuals, "consumers," industrial actors, whales, nations, coral reefs)? What are facts, what are fakes, what are concerns and controversies legitimate for debate and negotiations? Such questions are at the very core of the research methodology of Actor-Network Theory (Latour 2005). Latour have been pursuing such questions through many of his later works on climate change and sustainability in the anthropocene (Latour 2018).

Artificial intelligence (AI) is another example of a phenomenon of which the scale and lack of precedence transcend the established sciences' ability to fully comprehend or foresee the consequences. With the profound uncertainties with respect to how AI will develop and how it may change humankind and our living conditions, and not the least, how it will affect the global energy demand and thus tap directly and indirectly into the issue of climate change and nature loss, there is obviously—a bit ironic—much we do not know about the risks induced through the AI revolution. Because of the large uncertainties, AI represents a challenge for science and risk science. For example, the scientific method cannot be used to make accurate predictions. However, whether or not the scientific method is applicable is not decisive for the foundation of risk science.

Notably, public consensus on risk perception formed over time is fragile and influenced by processes of amplification and attenuation aided by sudden turns in political climate. Cultural characteristics relating to what humans value and prioritize are strongly affected by political trends and populist argumentation, and this is also a context in which science and research—including research on risk and uncertainty related to, e.g., climate change and AI—must constantly fight for legitimacy, making it difficult to project how these developments will affect our lives.

3.3 Implications and Conclusion

Risk science has a share in modernity. Its contributions to industrial development and societal wealth while ensuring safety along the whole spectrum from working environment to organizational accidents is evident for example in the "success" of the Norwegian petroleum history, where there has through the decades been a close collaboration between the risk community, the petroleum companies and the regulator. From being a field primarily occupied with organizational and industrial safety, risk science is now increasingly relevant and used for broader *societal* safety issues. With that comes a responsibility to contribute to developing economic sustainable industries and at the same time build knowledge on safety and risk for the benefit of not only the petroleum industry, but for risk science as a field as well.

When the tasks of the field are scaling up to serve not only the interests of organizations and industries, but also *societal and global* interests, it is increasingly important to acknowledge that although neutrality is the ideal of science, and should be fought for every day, science is politicized by many means. Hence, although we might like to think so, risk science does not more than other sciences produce objective, value-free facts, but knowledge understood as justified beliefs founded on data, information, tests, analysis, models, and theories, developed in a political context. This context may in different ways influence the scientific processes, for example through funding, negotiations, controversies, and alliances. Think for example about the fluctuating scientific trends/truths with respect to "root causes" of accidents. Where to look, and how far back causes should be traced, are issues of which scientists are not the only stakeholders. Facing giga-scale challenges like climate change and AI, whose effect on future living conditions are associated with enormous uncertainty, a critical risk science needs to be reflexive and avoid being unknowingly hijacked. Furthermore it needs to engage not only in resolving problems formulated and delineated by others, but to engage actively and critically with problem formulations, e.g., considerations associated with scaling. Choice of scale and boundaries always contain aspects of politics, but should not compromise the role of risk science which is to develop knowledge. For example, in relation to nuclear energy production, risk science reflexivity means to address all relevant risk and vulnerability aspects, also those that could challenge commercial interests.

Risk science has strong roots in engineering, but many central concepts have been internalized from interaction with other scientific fields (Antonsen et al. 2025; Haavik 2021). Strengthening the critical dimensions of risk science to meet giga-scale challenges with consequences for life conditions in societies with differing views on what are important values and what the world of tomorrow should look like, will require a further opening up of risk science to expand the intersection to other sciences, particularly the humanities and social sciences, to strengthen what has been labeled the research on risk science (Antonsen et al. 2025). A particular benefit of this will be that risk science can strengthen its ability to ask questions, besides to provide answers (or foundations for decisions). And, as we all know, in science as well as in politics, the questions we ask are often more important than the answers.

Acknowledgements For Torgeir Kolstø Haavik, the work has been funded by the Research Council of Norway, grant number 344332 (INTERSECT).

References

S. Antonsen, T. Aven, T.K. Haavik, Reflections on research in risk science and on risk science. Risk Anal., 1–14 (2025)

T. Aven, An emerging new risk analysis science: Foundations and implications. Risk Anal. **38**(5), 876–888 (2018)

T. Aven, *A History of Risk Science: Concepts and Principles of Risk Analysis* (Routledge, New York, 2025a)

T. Aven, On the relationship between disaster risk science and risk science. Risk Anal. (2025b)

T. Aven, O. Renn, On risk defined as an event where the outcome is uncertain. J. Risk Res. **12**(1), 1–11 (2009)

U. Beck, *Risk Society: Towards a New Modernity* (Sage, London, UK, 1992)

M.B. Brown, Politicizing science: Conceptions of politics in science and technology studies. Soc. Stud. Sci. **45**(1), 3–30 (2015)

U. Büntgen, The importance of distinguishing climate science from climate activism. Npj Clim. Action **3**(1), 36 (2024)

M. Douglas, A. Wildavsky, *Risk and Culture: An Essay on the Selection of Technological and Environmental Dangers* (University of California Press, 1983)

D. Easton, *The Political System: An Inquiry into the State of Political Science* (Alfred A. Knopf, New York, 1953)

P. Glasner, Reporting risks, problematising public participation and the human genome project, in *Environmental Risks and the Media*, (Routledge, 2013), pp. 130–141

S.T. Hansen, S. Antonsen, Taking connectedness seriously. A research agenda for holistic safety and security risk governance. Saf. Sci. **173**, 106436 (2024)

S.O. Hansson, T. Aven, Is risk analysis scientific? Risk Anal. **34**(7), 1173–1183 (2014)

T.K. Haavik, Debates and politics in safety science. Reliab. Eng. Syst. Saf. **210**, 1–7 (2021)

S. Jasanoff, The songlines of risk. Environ. Values **8**(2), 135–152 (1999)

B. Latour, *The Pasteurization of France* (Harvard University Press, Cambridge, MA, USA, 1988)

B. Latour, *Reassembling the Social: An Introduction to Actor-Network-Theory* (Oxford University Press, Oxford, UK, 2005)

B. Latour, *Down to Earth: Politics in the New Climatic Regime* (Wiley, 2018)

O. Renn, *Risk Governance: Coping with Uncertainty in a Complex World* (Routledge, 2008)

Chapter 4
Toward New Risk and Safety Imaginaries

Jean-Christophe Le Coze

Abstract This chapter develops an argument on the evolving landscape of safety and risks, characterized by a "global turn" in the social sciences. Ensuring AI safety in a worldwide-networked infrastructure or maintaining safe operating spaces within planetary boundaries affected by human activities are some of these challenges which come with our contemporary situation. It argues for form of systemic literacy shaped by study of events like COVID-19 and the use of analytical notions such as interconnectedness, synchronicity, nodes, flows, coupling, and complexity at the global scale for apprehending systemic and existential risks. These developments contribute to stimulate new risk and safety imaginaries for the safety management of socio-technical systems and infrastructures (e.g., natech, cybersecurity).

Keywords Systemic and existential risks · Anthropocene · Interconnectedness · Synchronicity · Nodes · Flows · COVID-19

4.1 Introduction

This chapter develops an argument on the evolving landscape of safety and risks, characterized by a "global turn" in the social sciences. This turn combines an incorporation of various phenomena such as globalization, digital expansion and the anthropocene. It leads to renewal of our risk and safety imaginaries, illustrated with COVID-19, and the introduction of vocabularies adapted to global events such as interconnectedness, flows, nodes, synchronicity, or systemic and existential risks.

J.-C. Le Coze (✉)
Ineris, Verneuil-en-Halatte, France
e-mail: Jean-Christophe.LECOZE@ineris.fr

C. Bieder et al. (eds.), *Living with the New Safety Landscape I*,
SpringerBriefs in Safety Management, https://doi.org/10.1007/978-3-032-29336-7_4

4.2 The Global Turn

The "*global turn*" in the social sciences at the end of the twentieth century and in the early twenty-first, addresses the fact that global realities affect societies in ways which cannot be ignored in studies any longer (Dufoix and Caillé 2013). The phenomenon of globalization became highly popular as well as contested in the 1990s following the end of the Cold War, with the fall of the Berlin wall in 1989. It corresponds to a moment in humanity history when the interconnections across continents intensified through a mix of technological possibilities (internet, air, and maritime transport) and political economy favorable to the idea of the free circulation of goods. It has opened a new chapter in world history following the collapse of USSR, the advent of China as a leading economy. The flows of goods, people, ideas, money, capital, images, information reshaped relationships between continents. Contributions from sociology, anthropology, history, geopolitics or geography helped frame this phenomenon, with heated debates on the relevance of the notion of globalization for understanding our contemporary situation, considering for instance that this process is not recent (e.g., Osterhammer 2015), or arguing on its negative (e.g., peripheries versus core economies—inequalities, exploitation, asymmetries) rather than its positive sides (e.g., Geiselberger 2017; Norberg 2018).

Global realities have also slowly become concretely evident from a planetary point of view through the phenomenon of global warming (and, before that, with the question of the ozone layer). By releasing carbon dioxide (CO_2) and methane (CH_4) in the atmosphere including with the development of these infrastructures supporting increasingly interconnected societies across the globe (Cowen 2014), infrastructures' functioning on fossil fuel, humanity has been heating the planet for decades after the second world war, in what is characterized as a great acceleration (McNeill and Engelke 2015). Other human-induced phenomena are also global such as the loss of biodiversity due to deforestation, resource depletion and chemically hostile environments for a diversity species (e.g., neonicotinoids for bees), or the eutrophication of the oceans because of the input in large quantities of nitrates (N) and phosphate (P) used in agriculture, and oceans' acidification as they absorb large quantities of the carbon released in the atmosphere. The notion of anthropocene captures this idea of our impact on the planet and on humanity in turn (Gemene and Rankovic 2019).

This characterization in two paragraphs of a "*global turn*" in the social sciences, for our description, conceptualization and understanding of societies, histories and of the human condition has only intensified in the past decades. The technologies driving the expansion of globalization, including the circulation of data through internet (cables, satellites) have only further developed in the past decade with the advent of social media and AI. The manifestation of global warming in recurring extreme events including heatwaves, megafires, storms or floods, one dimension of the anthropocene, have also intensified. They call for a radical revision of our established and core philosophies at the intersection of geology, biology, ecology, anthropology, sociology, politics, and history (e.g., Latour 2015); Clark and Szerszynski 2021; Chakrabarty 2023). Rising geopolitical tensions and global events, from COVID-19 to the Ukraine war (and its hybrid-warfare), trigger a recurring cycle of heated debates over the course of globalization, between its detractors who sees the beginning of its demise (e.g., Zeihan 2022), and those who remain its fervent defenders (e.g., Lincicome and Packard 2024).

4.3 Safety and the Global Turn

The field of safety is necessarily deeply affected by these trends. An indication of this global turn for this field is the explicit association and use of the notion of safety with new kinds of threats. Two examples have been selected from activities carried out by two organizations. The first is the concept of "*safe operating space*" developed by researchers at the Stockholm Resilience Centre, which studies the evolution of the human footprint on the planet in the Anthropocene era, with its associated risks, using the planetary boundaries (Richardson et al. 2023). The second example is "*Artificial Intelligence Safety*," from a document published by the AI Safety Institute based in San Francisco and their concerns with catastrophic risks generated by AI systems (Hendrycks et al. 2025).

The fact that safety becomes associated with some of the most pressing global issues and their related technological developments, the anthropocene and artificial intelligence, considered as representing potential of catastrophic risks, is an indication of a change of operating landscape, and the need for a renewal of our perspective in the field. Outside of this safety field as it is currently defined (e.g., Swuste et al. 2022; O'Hare 2023), in this global context, such reflections have begun. An author like Bridle argues for instance for the development of a form of "*systemic literacy*." He analyzes the deployment of a networked world, creating "*hyper-objects*" (a concept borrowed from Morton) that defy our understanding because of their sheer scale (Bridle 2018). Astonished by the unfolding realm of technological potentialities, his exploration of the new properties of the expanding range of algorithmically driven software, he is concerned with some of the latest developments such as high frequency trading, platform economies, states' surveillance, algorithmic management, smart bracelets and phones, predictive policing, algorithmic creativity in art … he calls for an ability to better grasp what is happening.

He elaborates, among many other things, on the positive feedback loops between the development of informational infrastructures (data centers) and global warming, such global scale phenomena for which we lack the ability to fully embrace them. Referring to Bridle, Voss develops a similar argument and elaborates on the need for new descriptive capabilities. Exploring the notions of systems, scale, legacy, matter, deviance, and breakage applied to several infrastructures embedded in the digital and algorithmic nature of these networks, she discusses logistics chains, aviation, architecture, cars, and finance, concluding that we are witnessing the expansion of unprecedented infrastructures (Voss 2024). The need to find words to describe these realities that defy our understanding is one crucial goal, considering that narratives are central to our grasp of these large-scale entities for their design and regulation. This selection of authors who tackle these issues indicates the wider concerns implied by the global, and our ability to understand it.

4.4 Toward New Risks and Safety Imaginaries

The title of this chapter, "*toward new risk and safety imaginaries*" contributes to this growing awareness of new scales of problems to be studied and conceptualized. In a series of articles and books, I have tempted to consider the implications of this "*global turn*" for the safety field (Le Coze 2018, 2020, 2023a, 2023b, 2024). Noting the increasing appearance of risk categories such as systemic and existential risks in the early years of the twenty-first century, I argue for their correspondence, respectively, with the intensification of globalization and associated events such as pandemics, terrorism or financial crises (systemic risks, OECD 2003), and the awareness of threats to civilizations through a range of events, from asteroid impacts, ecological crisis, bioterrorism to nuclear winter connected to planetary perspective (catastrophic/existential risks, Bostrom, and Ćirković 2008). These categories have since confirmed their status although with diverse orientations (e.g., Goldin and Muggah 2020; Ord 2020).

One interesting aspect for the safety field is that these new risk categories could be connected to the traditional, socio-technical risks (e.g., industries such as mining, nuclear, chemical, oil & gas, aviation, maritime), because of the centrality of such large technical systems, also now more commonly referred to as infrastructures, in the contemporary situation. These infrastructures are at the heart of the great acceleration—anthropocene (McNeill and Engelke 2015), and another expanding infrastructure, internet (i.e., cables, servers, see Blum 2019) is at the heart of the digital evolution, based on machine learning—AI (Bridle 2018). One way to explore the safety implications of this was to come back to the classic of the field, Normal Accidents by Perrow. Perrow was one of the authors who focused on such systems, from a risk perspective (Perrow 1984). To revisit the book to integrate the empirical changes and conceptual developments of the forty years since its publication in 1984 consists therefore in considering this "*global turn.*"

Post Normal Accident (Post NA) came as a description of this project of revisiting the classic. Post NA proceeds with five departures from the original formulation of NA (Le Coze 2020). This chapter focuses on the fifth, the expansion of the notion of coupling and complexity (the well-known categories of Perrow) at the global scale, toward the systemic literacy explicitly advocated by an author like (Bridle 2018), which also corresponds to a refoundation of our view of history, humanity, societies (e.g., Chakrabarty 2023). In this perspective, there is a strong interconnection between socio-technical systems (infrastructures) and globalization, but there is also an interconnection between globalization and existential/catastrophic risks (such as those linked to the anthropocene), which helps frame the category of global risks

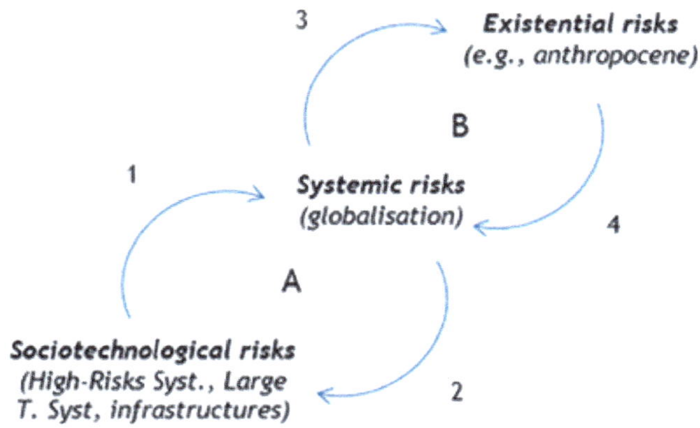

Fig. 4.1 Sociotechnical, systemic, and existential risks

(Le Coze 2023a). I have represented this in simple diagrams including the one below (Fig. 4.1) suggesting the importance of connecting these three categories because of their causal relationships. Those were developed in a text (a short version is reproduced here) based on the two loops of the diagram, A and B, and its numerated arrows (1–4) (Le Coze 2018, 2020).

Loop A. Causal Circularity Between High-Risk Systems and Globalization

1. High-risk, large technical systems are the infrastructures of our globalized world (e.g., ships, aircrafts, internet cables/servers, oil and gas). To study globalization and its associated systemic risks requires investigating the functioning and interactions of socio-technological systems at different scales, from national to supranational levels in complex networks and global value chains.
2. Globalization retroactively impacts these systems in at least three different ways. First, systemic risks represent threats to their operation (e.g., financial crisis, pandemics, terrorism); second, their degree of interconnectedness creates problems when one of these critical infrastructures is affected with ripple effects throughout the network. Third, globalization pushed by privatization, liberalization of finance and trade, ICT and transport revolutions has completely modified their operating conditions (as shown in the case of BP, or Boeing).

Loop B. Causal Circularity Between Globalization and the Anthropocene (and AI)

3. In relation to the anthropocene, globalization, has led to the development of many nations across the world and has become a key dynamic which contributes to global warming (one aspect of the anthropocene). Globalization is also a context of global competition creating incentives for nations to experiment technologies to deliver competitive and geopolitical advantages (e.g., artificial intelligence).

4. Finally, the anthropocene (e.g., global warming) represents direct threats also known as Natech as illustrated with the Fuskushima case (e.g., floods, hurricanes, heatwaves) for the functioning of infrastructures, and the anticipation of catastrophes has a considerable potential to limit the operations of large technical systems and infrastructures, if they were regulated. Potential catastrophic threats linked to artificial intelligence are also currently being discussed in the literature as introduced in AI Safety (see quote in Sect. 4.2).

These loops create a new unfolding operating landscape for safety-critical systems which alter their risk and safety profiles and the need of new risk and safety imaginaries to anticipate, prevent, mitigate, and prepare for surprises (e.g., natech, cybersecurity).

4.5 An Illustration with COVID-19

In the year of Post NA's publication in 2020, with this proposal to expand coupling and complexity to integrate the new magnitudes of impact of systemic/existential risks and their transnational regulatory challenges, the COVID-19 pandemics emerged. It thus became a clear illustration of this proposal, an example of global risk (Le Coze 2023a), a case of coupling and complexity at the global scale (Le Coze 2023b) connecting the anthropocene (existential risks), globalization (systemic risks) and infrastructures (sociotechnical risks) in the most explicit way imaginable. Following Guillèn (a co-author of Perrow in the 1990s), who applied the concepts of coupling and complexity to the global financial crisis of 2008 combined with notions of flows, nodes and their shock absorbing/diffusing properties (Guillèn 2015), it is interesting to highlight the networked interconnectedness of our contemporary situations at these different scales, at this global level of description (anthropocene, globalization, infrastructures, Fig. 4.2). Let's provide a brief description of this analysis, which is developed elsewhere more thoroughly (Le Coze 2023b).

To start with, the virus transmission from animals to humans is one outcome of deforestation, itself a product of the expansion of economic growth through globalization. By moving closer to wildlife, the probability of transmission of viruses to human is higher. The loss of biodiversity is also an explanation for a greater level of viruses' circulation, biodiversity acting as moderator, as a barrier against their multiplication (Robin 2021). This situation is therefore one expression of the anthropocene. Its existential threat is linked to viruses' lethality, and their transmission rate. In the COVID-19 case, remained a systemic event, not an existential one (the magnitude of impact was less than if the virus had been highly lethal and infectious).

The rapid spread of the virus (in hours and days) is a direct result of the infrastructures of our contemporary world enabling the circulation of various globalized flows, from aircrafts to ships, trains and cars in cities, acting as diffusing nodes of the virus. Cities are indeed some of the well-known connecting nodes of globalization. Responses by localities and states are therefore essential to understanding the shock

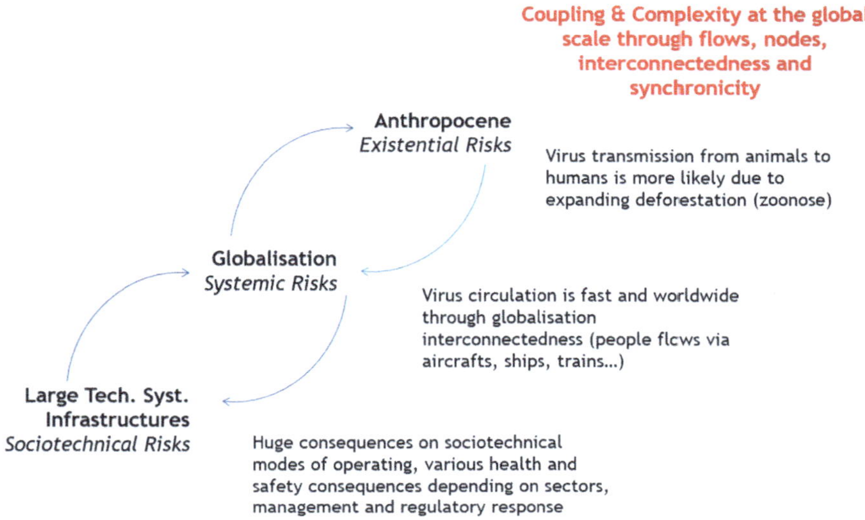

Fig. 4.2 COVID-19, coupling, complexity, interconnectedness, synchronicity

diffusing or shock absorbing properties of such nodes, as well as the systemic nature of the event (with their inequalities, asymmetries within and between nations) as well as transnational cooperation and governance. Decoupling measures were put in place, from re-establishing borders, grounding aircrafts to movement restrictions (i.e., lockdowns) in various places.

Finally, high-risk systems, large technical systems, and critical infrastructures had to re-configure their modes of operating, through resilient and adaptive strategies of actors, organizations, regulations, and institutions, testing their ability to cope in unknown circumstances, including the ability to quickly find a vaccine. New ways of working, including digital solutions such as working from home on a massive scale created such adaptations, and various innovations were invented on the ground, challenging the traditional hierarchies, and top-down modes of control to favor more flexible strategies. This constitutes an unprecedented experience at the global scale, everywhere in the world.

To this very brief description of COVID-19 highlighting the interactions at different scales, another key notion is needed to characterize this expanding global landscape, that of synchronicity, namely the very rapid causal relationships across continents created by this increased interconnectedness of flows and networks. Developed by global historians to emphasize the peculiarities of our historical phase of increased density of networks and systems linking many different regions of the world, this concept applies quite well to many global events (Conrad 2016).

4.6 Conclusion

This chapter develops an argument on the evolving landscape of safety and risks, characterized by a "global turn" in the social sciences. Ensuring AI safety in a worldwide-networked infrastructure or maintaining safe operating spaces within planetary boundaries affected by human activities are some of these challenges which come with our contemporary situation. A form of systemic literacy shaped by study of events like COVID-19 and the use of analytical notions such as interconnectedness, synchronicity, nodes, flows, coupling and complexity at the global scale for apprehending systemic and existential risks. These developments contribute to stimulate new risk and safety imaginaries for the safety management of socio-technical systems and infrastructures (e.g., natech, cybersecurity).

References

A. Blum, Tubes: A Journey to the Centre of the Internet. Ecco. (2019)
N. Bostrom, M.M. Ćirković (eds.), *Global Catastrophic Risks* (Oxford University Press, Oxford, 2008)
J. Bridle, *New Dark Age* (Technology and the end of the future, London, Verso, 2018)
D. Chakrabarty, *Après le changement climatique* (Changer l'histoire, Paris, Gallimard, 2023)
N. Clark, B. Szerszynski, *Planetary Social Thoughts. The anthropocene challenge to the social sciences* (Polity Press, London, Cambridge, 2021)
S. Conrad, What is global history? Princeton: Princeton University Press. (2016)
Cowen, *The Deadly Life of Logistics* (Mapping Violence in Global Trade, 2014)
S. Dufoix, A. Caillé, *Le tournant global des sciences sociales* (Paris, La Découverte, 2013 dir). [The social science global turn]
H. Geiselberger (ed.), *L'âge de la régression* (Paris, Folio Essais, 2017)
F. Gemene, A. Rankovic, *Atlas de l'anthropocène* (Presses de Science Po, 2019)
I. Goldin, R. Muggah, *Terra Incognita: 100 Maps to Survive the Next 100 Years* (Penguin Book, 2020)
M. Guillén, The architecture of collapse. The Global System in the 21st Century. Oxford: Oxford University Press. (2015)
D. Hendrycks, M. Mazeika, T. Woodside, An overview of catastrophic AI risks. Report retrieved from https://arxiv.org/pdf/2306.12001 in March 2026 (2025)
Latour, *Face à Gaïa* (La Découverte, Paris, 2015)
J.C. Le Coze, An essay: Societal safety and the global 1,2,3. Saf. Sci. **117. Part C**, 23–30 (2018)
J.C. Le Coze, *Post Normal Accident. Revisiting Perrow's Classic* (CRC Press. Taylor and Francis, Boca Raton, FL, 2020)
J.C. Le Coze, Global risks, in *Handbook of the Sociology of Globalisation*, ed. by D. Hofäcker, C. Karner, (Edward Elgar, 2023a)
J.C. Le Coze, Coupling and complexity at the global scale. Flows, networks, interconnectedness and synchronicity (e.g., COVID-19). Saf. Sci. **106193** (2023b)
J.C. Le Coze, Climate change, global scales and safety, in *Climate Change and Safety in High-Risk Industries*, ed. by C. Bieder, G. Grote, J. Weyer, (SpringerBriefs in Applied Sciences and Technology, Springer, Cham, 2024)
S. Lincicome, C. Packard, *Defending Globalization: Facts and Myths about the Global Economy and its Fundamental Humanity* (Cato Institute, 2024)

J.R. McNeill, P. Engelke, *The Great Acceleration: An Environmental History of the Anthropocene since 1945* (Harvard University Press, Cambridge, 2015)
J. Norberg, Non, ce n'était pas mieux avant. 10 bonnes raisons d'avoir confiance en l'avenir. Pocket. (2018)
O'Hare, *Introduction to Safety Science* (Routledge, People, Organisations and Systems, 2023)
T. Ord, *The Precipice* (Grand Central Publishing, Existential risk and the survival of humanity, 2020)
Organization for Economic Development and Cooperation. *Emerging systemic risks: final report to the OECD future proeject* (Paris, france: OECD, 2003)
J. Osterhammer, *The Transformation of the World: A Global History of the Nineteenth Century* (Princeton University Press, Princeton, 2015)
C. Perrow, Normal Accidents. Living with high-risk technologies (New York: Basic Books, 1984)
M. Robin, *La fabrique des pandémies* (Préserver la biodiversité un impératif pour la santé planétaire, Le Découverte, 2021)
J. Richardson, W. Steffen, W. Lucht, J. Bendtsen, S.E. Cornell, J.F. Donges, I. Fetzer, et al., Earth beyond six of nine planetary boundaries. Sci. Adv. **9**, 37 (2023)
Swuste et al., *From Safety to Safety Science* (Routledge, The evolution of thinking and practice, 2022)
Voss, *System Ultra. Making sense of technology in a complex world* (London, Verso, 2024)
Zeihan, *The End of the World Is Just the Beginning* (Mapping the collapse of globalization, Harper Business, 2022)

Chapter 5
AI Gold Rush 2 Versus AI Realism: Overplaying Envisioned Benefits, Underplaying Real Risks

Michael F. Rayo, Dane A. Morey, Tor Olav Grøtan, and David D. Woods

Abstract The current Gold Rush to seize advantage from AI-infused technologies clashes with modern safety management. These new technologies are mistake-prone, yet the rush for advantage is producing enormous pressure for deployment in every sector, including safety-critical settings. This chapter examines the contrast between the *rhetoric* of AI Gold Rush 2 and the *limits* of the technologies based on findings from past technology gold rushes, including AI Gold Rush 1 in the 1980s. The contrast reveals patterns of growth, complexification, and adaptation as new capabilities transform human-technological systems in unexpected ways. These patterns provide organizations with a pragmatic perspective to navigate the turbulence and instabilities of technology-induced change: *AI Realism* grounded in the foundations of Resilience Engineering and Joint Cognitive Systems.

Keywords AI realism · AI Gold Rush · AI misbehavior · Joint activity · Human–AI systems · Safety-critical systems

5.1 On the Front Lines of a Tech Gold Rush

Safety-critical industries, and safety professionals within them, are bombarded by proposed deployments of artificial intelligence (AI) algorithms. Many of these algorithms, including Generative AI and Large Language Models (LLMs), appear easily

M. F. Rayo (✉) · D. A. Morey · D. D. Woods
The Ohio State University, Columbus, OH, USA
e-mail: rayo.3@osu.edu

D. A. Morey
e-mail: morey.38@osu.edu

D. D. Woods
e-mail: woods.2@osu.edu

T. O. Grøtan
Sintef, Trondheim, Norway
e-mail: tor.o.grotan@sintef.no

C. Bieder et al. (eds.), *Living with the New Safety Landscape I*,
SpringerBriefs in Safety Management, https://doi.org/10.1007/978-3-032-29336-7_5

accessible to use in a wide array of tasks due to their natural language interaction, responsiveness to human instructions, and apparent generality. However, are these AI-infused technologies ready for the real risks of safety-critical operational worlds? What new work and costs are necessary to deploy AI-infused systems safely and effectively into dynamic risky worlds? Though the rhetoric surrounding these technologies promise powerful and quickly implementable capabilities, the reality of rushing to AI for advantage can be quite different.

A Safety Manager's Story of the AI Gold Rush:

> *Your boss is asking you and every other manager: "What are you doing to utilize AI?" It's hard to believe how quickly things have changed. Chatbots and LLMs were another futuristic news story just months ago. Now these technologies seem to be everywhere! You've heard people comparing AI to the creation of the internet, the industrial revolution, and even the invention of papyrus. To hear them talk, the history is already written. AI is the next paradigm shift—only those who adopt it will survive.*
>
> *AI is now* the *priority, coming from the top of your organization. Executives believe in the promise of AI to not only work faster, cheaper, and better, but also as a necessity to maintain competitive advantage. "Our competitors are already integrating AI into their operations," your CEO declares. "They're increasing productivity and reducing costs. We all must embrace AI quickly or we'll be left behind."*
>
> *As head of safety, you are responsible to sustain a record of reliability, minimize accidents, investigate incidents, implement changes, and document operations to industry and professional standards. Incidents and accidents have consequences—loss of valued services, financial losses, worker injuries, and threats to public safety. Because the safety department interacts with every aspect of operations, you have a unique vantage point to observe how the integration of AI capabilities is changing the ways your organization operates over time.*
>
> *As you look at your organization, you are surprised to find that many have already been using AI-infused technologies in their work! The accessibility of LLM chatbots requires no special training or equipment, making it easy for people to use them without following the usual implementation processes. For some, these technologies seem nice to have, simplifying or expediting some nominal tasks which remain easy to accomplish without AI. For others, LLMs have been adopted as the de facto method to conduct critical work more efficiently. AI-infused work is quickly becoming the new norm. As upper management continues to expect more work with less staff, AI is increasingly critical to the business of providing valued services. Staff are no longer able to accomplish everything without using AI. This became painfully clear when operations ground to a halt during a recent outage from a major web service provider which took down a popular chatbot. Looking closer, you notice the previous*

> *methods, skills, and tools for work are visibly decaying. As AI becomes increasingly central to your organization, you have a frightening realization: AI is now used only one step away from safety-critical activities, and it's happening without the engagement of the safety department.*
>
> *In meetings, upper management assuages your concerns about the erosion of safety, including the lack of safety engineering and testing. "These technologies greatly improve our productivity," they say, "It's your job to make sure they also improve safety." Management sees productivity gains and the opportunity to lower costs, especially by reducing the need for experienced (and expensive) personnel. In the rush to gain advantage from AI, there is no time for the slow process of reviewing, testing, and checking new systems for offsetting costs or risks. LLMs can be implemented with minimal development effort, arriving pretrained on massive data sets to provide capabilities through natural language interaction. To seize advantage quickly, upper management begins cutting staffing across the organization, including in the safety department. "Using LLMs can offload most of your documentation and reporting," they explain, "so each member of the safety staff can cover a larger portion of the organization." With suddenly more work and less staff, you feel you don't have much of a choice.*
>
> *Retreating to your office, you feel trapped. It seems like your organization is hit not simply by a new technology but by a new strategy—the Silicon Valley mantra of "move fast and break things," or even worse, "break things to move fast." Since when has your organization prioritized rapid change over evidence-based engineering, safety, and decision-making? The intense pressure to take advantage of AI seems at odds with the proactive safety program your department has worked so hard to infuse throughout the organization—especially because these technologies are mistake-prone. You worry that management does not fully understand the new risks that could be introduced by these changes, but you also know that slowing the pace of change without hard evidence will enrage upper management. You fear it is only a matter of time before something goes wrong and you'll be held responsible for the harm, both legally and morally. What can you do?*

The safety manager in this example is trapped in an authority-responsibility double-bind: responsible for safe operations without the resources to carry out that responsibility. The rush to capitalize on the anticipated benefits of AI encourages fast implementations without thoroughly checking for offsetting risks or costs. Stuck in the middle, the safety manager is squeezed from all sides, constricted by:

- a *top-down* pressure from management to use AI for cost reduction;
- a *bottom-up* pressure from front-line workers integrating AI in ways that are increasingly critical to operations;
- an *outside-in* pressure from competitors fueling narratives that adopting AI is critical to maintaining competitive advantage;

- an *inside-out* pressure from staffing cuts necessitating AI to cope with overload.

Each of these pressures for rapid change are potential destabilizing forces which can undermine the processes for assuring safe and effective systems, incentivizing cutting corners and taking shortcuts to gain advantage. In the rush for advantage, prudent engineering practices are ignored or sidestepped in the quest for speed. This can be seen in the engineering and regulatory decisions leading up to the two Boeing 737 Max accidents that killed 346 people (Dekker and Woods 2024). To strike a balance between the advantage of moving fast and the need to check how changes can produce new vulnerabilities, safety managers and their organizations will need to recognize early signals that trigger more thorough investigation. In this chapter, we highlight some patterns of instability and vulnerability from both past and current AI Gold Rushes to guide the safe and effective deployment of AI-infused technologies.

5.2 Enduring Patterns from Past AI Gold Rushes

Technology gold rushes are *human* stories of seeking future advantage when new technological capabilities appear. Though the specific technologies differ, the cycles of growth, complexification, and adaptation reveal general patterns that are directly applicable to the current wave of AI-infused technologies.

5.2.1 Overconfident and Underspecified Claims

One hallmark of a technology gold rush is a gap between claims about a technology and the actual challenges that arise as the technology is deployed. Highly controlled demonstrations of AI algorithms produce optimism and an increasingly distorted view of actual capabilities. This gap between claims and actual capabilities was exemplified in the rush to develop and deploy expert systems in the 1980s, or *AI Gold Rush 1*. The claims about these technologies were found to (1) underspecify the challenges of integrating AI into complex and risky worlds of practice, (2) overestimate expected benefits, (3) and underestimate actual costs, vulnerabilities, risks, and failures (Woods and Dekker 2000). Evidence contradicting gold rush rhetoric was often discounted.

5.2.2 No AI is Autonomous in Context

Algorithmic technologies cannot *independently* perform cognitive work in context. All work in AI-infused systems is performed *jointly* between humans and machines, *distributed* over multiple human/machine roles, and *situated* in complex settings with extensive interdependencies (Hutchins 1995). This in part stems from the fact that all technologies are limited and fallible, yet people remain responsible for system outcomes. It is therefore necessary and advantageous to support the *interdependence* of people and technologies (Johnson and Vera 2019). Deprioritizing human involvement does not lessen these interdependencies but instead removes supports, resulting in a poor joint human–AI system.

5.2.3 *Complexification of Failure*

Technology failures often spur the development of additional technologies to monitor, restrict, or safeguard other technologies. However, safeguards are no less fallible than the technologies they monitor. The failure of monitoring automation can create convoluted anomalies that are incredibly difficult for people to diagnose within the tempo of operations. For example, the automation intended to monitor for faulty sensor data onboard Turkish Airlines Flight 1951 failed, allowing erroneous data to be used by other automated systems. This triggered a cascade of other failures which further obscured the failure of the monitoring automation. Therefore, using fallible automation to monitor fallible automation does not necessarily reduce the frequency or severity of failures, but rather *shifts* the kinds of failures that occur. This creates new and increasing demands for people to monitor the automation which monitors other automation.

5.2.4 *Envisioned Benefits are Offset by Real Costs*

New deployments of automated capabilities anticipate large benefits for little cost after initial investments are applied at scale. However, cycles of growth, complexification, and adaptation reveal a variety of *offsetting costs* as "new forms of conflict, congestion and cascade arise as apparent benefits are hijacked" (Woods 2020, p. 101). These costs include new demands of work, new risks to manage, and new kinds of anomalies. Often, this exposure to new or modified risks is hidden and only discovered *after* deploying new technologies. The full costs of deploying technologies are not reduced but rather deferred in time or transferred to others.

5.3 New Reverberations in AI Gold Rush 2

The claims surrounding this latest wave of AI-infused technologies, or *AI Gold Rush 2*, are more public, impassioned, and diverse over a wider range of people than any previous Gold Rush. The claims and counterclaims differ to the maximum degree possible, ranging from accelerationist visions of utopian abundance to fatalist visions of apocalyptic catastrophe. These noisy, overcrowded, and wildly contradicting views make it difficult to discern fantasy from reality.

Additionally, AI Gold Rush 2 technologies have some consequential technical differences from those in AI Gold Rush 1. First, there has been a shift from deductive models to inductive (or statistical) models, driven by machine learning techniques. Second, current technologies are much more accessible and interconnected than their predecessors. We examine some of the new challenges these technologies produce through the lens of a seemingly ubiquitous AI family: Large Language Models (LLMs).

5.3.1 From Clumsy to Sloppy

Traditional automation was described as "clumsy" because it alleviated workload in normal situations but added to workload in challenging situations. Generative AI technologies often add to workload even in normal situations. The enormous volume of low-quality outputs produced by LLMs is colloquially referred to as "AI slop." Although it is fast and easy for these algorithms to generate a variety of outputs that appear to be promising first steps, it is enormously difficult for people to filter and revise outputs toward an acceptable product. This reveals new offsetting costs. Productivity gains are offset by the costs of filtering and revising, which sometimes takes longer than if people had worked without AI (Becker et al. 2025). Generating high-quality outputs also requires the development of new human expertise, like "prompt engineering." Contrary to gold rush rhetoric, LLMs of the future could be even *sloppier*, as the current sloppy outputs of generative AI models progressively pollute the digital databases which are used to train future models.

5.3.2 From Interrogable to Inscrutable

The shift toward larger and more complex AI models has increased the difficulty of understanding and recognizing AI limits. In AI Gold Rush 1, models often reflected recognizable structures of the real world. In AI Gold Rush 2, the shift toward large statistical models (e.g., neural networks) has come at the cost of recognizable structures. Although the computations of these models are technically feasible to extract, the lack of understandable structure renders them inscrutable. Even their developers cannot fully comprehend how outputs are computed from inputs. These models remain fallible and are especially prone to *out-of-the-blue errors*, or perplexing mistakes in situations similar to those that produce successes (Kim et al. 2025). These errors are a symptom of poorly understood capability limits. Explanations of AI outputs are an incomplete solution because they too are persuasive and fallible, which can reinforce erroneous AI outputs or introduce additional sources of error

(Morey et al. 2025a). There remains little evidence to suggest how or if these techniques for revealing AI computations can be leveraged to help people recognize and recover from AI errors.

5.3.3 From Literal-Minded to Generative

With traditional automation, most mistakes could be attributed to the literal-mindedness of machines whose underlying models were mismatched with the world. These behaviors could be surprising in the moment, but were always understandable in retrospect. In contrast, the generative nature of LLM outputs means that errors are also generative. There are no guarantees that the statistical construction of outputs from LLMs will be factually accurate or semantically meaningful. These generative errors of commission have been anthropomorphized as "hallucinations,"[1] but they are simply a result of statistically constructed outputs. These errors are inherent to all generative AI algorithms (Kim et al. 2025).

5.3.4 From Incoherent to Plausible

Compounding the previous issues, the outputs of AI-infused technologies are increasingly believable, even when they are not accurate. Algorithms based on machine learning are trained to output the most statistically plausible answers. For LLMs, this results in impeccable syntax and reasonable logic which indeed appears to produce trustworthy outputs. However, these technologies remain fallible and mistake-prone (Kim et al. 2025). Plausible AI recommendations increase the difficulty for people to recognize and recover from errors. It requires greater human expertise to discard plausible yet incorrect outputs, which contradicts the promise of pairing AI with less experienced workers. Compounded by the aforementioned inscrutability, *plausible-yet-fallible* AI creates conditions ripe for authority-responsibility double-binds, where people remain responsible for outputs but are undermined in their ability to recognize and recover from AI errors.

5.3.5 From Limited to Illusory

The human-like familiarity of natural language interactions with LLMs has another insidious risk: the *illusion of joint system capabilities*. People are remarkably quick to attribute human-like intentions to simple machine behaviors. Technologies which mimic human-like outputs will only amplify this tendency. Personification

---

[1] For a compilation of AI hallucinations, see: https://www.damiencharlotin.com/hallucinations/

of LLMs risks creating the appearance of human-like capabilities while masking well-established vulnerabilities. For example, LLMs are notoriously overconfident, quick to recommend, poorly calibrated to concede uncertainty, yet also quick to agree during conflict. These algorithms are limited in evaluating their own output, leaving people to assume the burden of detecting AI limitations. However, LLMs may misleadingly encourage people to look for human-like cues of miscalibration when the actual cues will be different, undermining the effectiveness of people to cross-check LLM outputs. Conversely, LLM cross-checks of people appear to be effective because of their natural language abilities, but mask LLM biases for sycophancy. Effective joint activity requires individuals to convey signals of miscalibration *to others* (and for others to act upon those signals) even when they are unable to detect their own miscalibration (Jalaeian et al. 2025). Human-like LLM outputs risk overestimating joint system capabilities and underestimating risks.

5.3.6 From Local to Global Interdependencies

Because of the widespread use of AI and dependence on digital infrastructure for data storage and computation, local performance is now increasingly dependent on a global set of capabilities and actors. This increases cybersecurity risks across these extended critical networks and the need for cyber-resilience practices (Pettersen and Grøtan 2024). However, current cybersecurity practices often utilize highly automated processes to detect and correct problems, which are themselves fallible and subject to the preceding limitations. The increasing use of AI to generate computer code amplifies these risks, especially for long-term software maintenance (Chen et al. 2026). Recent critical digital services outages[2] have illustrated how automation meant to mitigate failures can also trigger and complexify failures. Thus, the increased (and necessary) attention to AI security must also be designed as a joint human–AI cognitive system.

5.4 Safety's Contribution: Focusing on Challenges

Understanding how systems respond to the most challenging foreseen and unforeseen situations is the hallmark of modern Resilience Engineering (Hollnagel et al. 2006). As complex systems reach or extend past their operating envelopes, how systems adjust and recruit new resources to support joint activity is critical for sustaining performance (Woods 2018). Because gold rushes discount or ignore many of these

[2] For example, see the Amazon Web Services (AWS) outage on October 20, 2025, where automation intended to monitor and correct faults instead propagated a fault, leading to a cascade of failures and widespread outages across a variety of services: https://aws.amazon.com/message/101925/

challenges, it is particularly important for Safety to emphasize high-stress situations and legitimize sustained investment in joint and adaptive capabilities.

5.5 The Way Forward: Humble but Different Engineering

Although it is tempting to be overwhelmed by the excitement of these new technologies, it is dangerous to over-project or overestimate future capabilities based on successes in early, more controlled demonstrations (Chen et al. 2026). Humble engineering requires that capabilities are never assumed but rather evaluated and supported by corresponding evidence. We offer three specific points for what these evaluations must encompass.

5.5.1 Monitor for Vulnerabilities

Methods must address a variety of challenges that create vulnerabilities in complex adaptive systems. These challenges include those associated with abstract characteristics of cognitive work (Woods and Hollnagel 2006) and also specific vulnerabilities known to a class of technologies (as mentioned above). Systems must have mechanisms to monitor for expected vulnerabilities and detect when new or unexpected vulnerabilities arise.

5.5.2 Explore Future Potential Offsetting Costs

The complexities arising from today's successes will drive tomorrow's challenges (Woods 2020). These, too, have recurring patterns (Woods and Hollnagel 2006). Initiatives that reduce human personnel and human expertise will reduce future adaptability. Initiatives that improve throughput and efficiency also typically introduce increased coupling, congestion, and conflicts that are difficult to fully resolve. Responsible systems engineering requires that these future potential offsetting costs are accounted for in their evaluation of candidate solutions.

5.5.3 Test Jointly, Situated, Early, and Often

Responsible testing of AI-infused technologies must include a balanced representation of good, mediocre, and poor AI performance (Morey et al. 2025a). Even infrequent poor system performance requires attention because of the potential severe consequences in safety-critical systems. Testing must also empirically measure the performance of people and AI together (Morey et al. 2025a). Isolated performance in laboratory environments (e.g., benchmarking) doesn't reliably predict operational performance when technologies are working jointly with people. Instead, testing strategies that evaluate systems jointly across a range of AI performance have provided valuable insights into vulnerabilities that can be remedied before deployment and resultant harm (Morey et al. 2025b).

Even promising human-in-the-loop (or human-on-the-loop) predeployment evaluations do not guarantee that technologies will work in production or sustainably work as the world (and technology) changes. Responsible AI testing must also involve methods for early and ongoing evaluations that continue throughout the technology lifecycle. Design heuristics and other analytic evaluations can be used early by designers as inexpensive ways to detect and address downstream limitations (Jalaeian et al. 2025). Continuous assurance methods provide lightweight analysis of systems-level signals in the operational environment, allowing for periodic monitoring that can drive proactive interventions (Morey et al. 2022).

5.6 Conclusion: The Groundwork for AI Realism

In this chapter, we mapped out a third perspective on the AI Gold Rush: AI Realism. This offers a pragmatic and balanced alternative to the rhetoric projecting either doom or abundance. The latter races for envisioned benefits while the former attempts to rein in the rush. The result is instability. In contrast, AI Realism is based on results from past cycles of technology change where people seeking advantage adapted capabilities in surprising ways, new vulnerabilities and complexities accompanied growth, and some human roles adapted to cope with new complexities. Finding and defusing the new situations that present challenge and surprise requires humility and thorough testing in safety systems engineering. Discovering how to use new capabilities to tackle hard problems that plague today's systems requires partnerships between problem holders, stakeholders, and technologists. These efforts are necessary and add costs or delays relative to the naïve allure of cheaper, faster, better results in dynamic risky systems. When the pressure to "move fast and break things" is strong, testing and design shortcuts produce complexity penalties that are difficult and costly to manage after algorithms have been deployed. We have seen this story before: it resulted in fatal accidents and outages of critical and valued services.

In this cycle, the proclivity for technology to make mistakes and to accumulate slop creates new, dramatic risks. AI Realism balances finding opportunities and mitigating vulnerabilities, but only if safety systems engineering evolves and broadens its scope and methods. The new science and engineering for resilient performance already begins to show the way to design for synchronized joint activity, to support cognitive work carried out by diverse sets of human and algorithmic roles, and to be rigorous and pragmatic in new ways, at new speeds, and encompassing new scales (Herrera et al. 2024). In this age of Actual Instability (AI), anything else would amount to safety and systems engineering malpractice.

References

J. Becker, N. Rush, E. Barnes, D. Rein, Measuring the impact of early-2025 AI on experienced open-source developer productivity. https://arxiv.org/abs/2507.09089). arXiv. (2025). https://doi.org/10.48550/arXiv.2507.09089

J. Chen, X. Xu, H. Wei, C. Chen, B. Zhao, SWE-CI: evaluating agent capabilities in maintaining codebases via continuous integration. arXiv. Preprint posted online March 4, 2026. https://doi.org/10.48550/arXiv.2603.03823

S.W.A. Dekker, D.D. Woods, Wrong, strong, and silent: what happens when automated systems with high autonomy and high authority misbehave? J. Cognitive Eng. Decis. Making **18**(4), 339–345 (2024). https://doi.org/10.1177/15553434241240849

S.W.A. Dekker, M.D. Layson, D.D. Woods, Repentance as rebuke: betrayal and moral injury in safety engineering. Sci. Eng. Ethics **28**(6), 56 (2022). https://doi.org/10.1007/s11948-022-00412-2

I. Herrera, R. Patriaca, A. Adriaensen, M. Rayo, E. Rigaud, D.D. Woods, E. Lay, P. Ferreira, L. Maguire, L. David, C. Nemeth, J. Lundberg, J. Franca, S. Chuang, C. Alexander, R. Woltjer, J. Allspaw, N. McDonald, B. Johansson, *Resilience Engineering 20 Years—Progress, Challenges and Opportunities—Notes from Porto Meeting* (Zenodo, 2024). https://doi.org/10.5281/zenodo.14507214

E. Hollnagel, D.D. Woods, N. Leveson (eds.), *Resilience Engineering: Concepts and Precepts* (Ashgate Publishing Company, 2006)

E. Hutchins, *Cognition in the Wild* (Bradford Books, 1995)

M. Jalaeian, D.A. Morey, M.F. Rayo, Joint activity design heuristics for enhancing human-machine collaboration (no. https://arxiv.org/abs/2512.08036). arXiv. (2025). https://doi.org/10.48550/arXiv.2512.08036

M. Johnson, A. Vera, No AI is an Island: the case for teaming intelligence. AI Mag. **40**(1), 1 (2019). https://doi.org/10.1609/aimag.v40i1.2842

Y. Kim, H. Jeong, S. Chen, S.S. Li, M. Lu, K. Alhamoud, J. Mun, C. Grau, M. Jung, R. Gameiro, L. Fan, E. Park, T. Lin, J. Yoon, W. Yoon, M. Sap, Y. Tsvetkov, P. Liang, X. Xu, et al., Medical hallucination in foundation models and their impact on healthcare (p. 2025.02.28.25323115). medRxiv. (2025). https://doi.org/10.1101/2025.02.28.25323115

D.A. Morey, M.F. Rayo, M. Li, From reactive to proactive safety: joint activity monitoring for infection prevention. Proc. Int. Symp. Hum. Factors Ergon. Healthc. **11**, 48–52 (2022). https://doi.org/10.1177/2327857922111009

D.A. Morey, M.F. Rayo, D.D. Woods, Empirically derived evaluation requirements for responsible deployments of AI in safety-critical settings. Npj Digit. Med. **8**(1), 374 (2025a). https://doi.org/10.1038/s41746-025-01784-y

D.A. Morey, M.F. Rayo, D.D. Woods, How AI can degrade human performance in high-stakes settings. AI Front. (2025b) https://ai-frontiers.org/articles/how-ai-can-degrade-human-performance-in-high-stakes-settings

S. Pettersen, T.O. Grøtan, Exploring the grounds for cyber resilience in the hyper-connected oil and gas industry. Saf. Sci. **171**, 106384 (2024). https://doi.org/10.1016/j.ssci.2023.106384

D.D. Woods, The theory of graceful extensibility: basic rules that govern adaptive systems. Environ. Syst. Decis. **38**(4), 433–457 (2018). https://doi.org/10.1007/s10669-018-9708-3

D.D. Woods, The strategic agility gap: how organizations are slow and stale to adapt in turbulent worlds, in *Human and Organisational Factors: Practices and Strategies for a Changing World*, ed. by B. Journé, H. Laroche, C. Bieder, C. Gilbert, (Springer International Publishing, 2020), pp. 95–104. https://doi.org/10.1007/978-3-030-25639-5_11

D.D. Woods, S. Dekker, Anticipating the effects of technological change: a new era of dynamics for human factors. Theor. Issues Ergon. Sci. **1**(3), 272–282 (2000). https://doi.org/10.1080/14639220110037452

D.D. Woods, E. Hollnagel, *Joint Cognitive Systems: Patterns in Cognitive Systems Engineering* (CRC Press, 2006). https://doi.org/10.1201/9781420005684

Chapter 6
Toward a New Social Contract for Safety

Gudela Grote

Abstract The concept of the social contract has been introduced in political philosophy to capture agreements between the ruled and their rulers, but it is also used to describe exchange relationships between social actors more broadly. In this chapter I argue that societies can be better protected from harm by understanding and actively shaping social contracts between the firms whose business entails safety risks, the public including the firms' employees who are subjected to these risks, and governmental and other institutions which oversee firms' risk management to protect the public. The social contract for safety has been breached in several ways, especially regarding the promise of full risk control through technology and business interests being kept in check by safety concerns. To establish a new contract for safety, technocratic assumptions need to be overcome and a new order of risk management established. I postulate and illustrate three fundamental principles to renew the social contract for safety: control–accountability alignment, decentralized governance, and integrative stakeholder negotiations. Strong alliances among international institutions, such as regulatory bodies from various industry sectors and the ILO, are needed to push for action. Having a broad international outlook on the path forward will foster consideration of the diversity of working and living conditions across the globe, hopefully preventing a further separation between the haves and have nots in this world.

Keywords Social contract · Risk and safety management · Contract breach · Stakeholder governance · International alliances

In the most general terms, a contract defines agreed-upon rights and obligations of two or more parties. The concept of the social contract has been introduced in political philosophy to capture agreements between the ruled and their rulers, where individuals give up some of their freedom in exchange for protection of their remaining rights (Encyclopedia Britannica 2025). However, it is also used to describe exchange

G. Grote (✉)
ETH Zürich, Zurich, Switzerland
e-mail: ggrote@ethz.ch

C. Bieder et al. (eds.), *Living with the New Safety Landscape I*,
SpringerBriefs in Safety Management, https://doi.org/10.1007/978-3-032-29336-7_6

relationships between social actors more broadly, such as between employers and employees or their respective representatives. In the present context of risk and safety management, the social contract concerns relations between firms whose business entails risks for health, safety, or the environment, members of the public including the firms' employees who are subjected to these risks, and governmental and other institutions which oversee firms' risk management to protect the public. I will argue in the following that to understand such social contracts and how they may have to change to better protect societies from harm, the underlying exchange mechanisms need to be studied including implicit assumptions and expectations by all parties that are captured by the concept of a psychological contract. I will then postulate and illustrate three fundamental principles to renew the social contract for safety.

6.1 Social Exchange

Any social relationship entails social exchange, that is the give and take between the parties in the relationship. This exchange can be defined in economic terms, such as receiving certain goods or services for a certain amount of money, and may be legally enforceable. It can also be more vague and comprise expectations about material or immaterial benefits to be gained from the exchange. Key to any social exchange is the norm of reciprocity (Gouldner 1960), that is the assumption that positive acts will be responded to by positive acts and negative acts by negative ones. Reciprocity does not necessarily require identical things to be exchanged but can be tailored to the specific needs of each party. The norm of reciprocity strengthens a sense of obligation among the parties and helps to build trust. These effects will be stronger if there are little self-interest and external pressure involved and if the receiving party is in great need and the giving party has to exert considerable effort.

Next to reciprocation, social exchange is also experienced more positively when it is perceived as just. It is important to note that justice does not only concern the distribution of material and immaterial goods between the involved parties, but also the procedures used to decide on the distribution and transparency of these procedures (Lind and Tyler 1988). An unequal distribution of benefits—or harms for that matter—may still be perceived as just if the process by which the decision was taken is transparent and acceptable to all parties.

6.2 Psychological Contracts

The less formalized social exchange relationships are, the more equivocal and implicit the obligations become that govern the exchange. Implicit reciprocal expectations about the give and take between parties have been termed psychological contracts (Rousseau 1989). However, even formalized relationships defined by legal contracts, such as employment relationships, will always also contain implicit expectations that potentially may override what has been formally agreed. For instance, an employee with a fixed-term contract may become convinced that their contract will be prolonged or turned into a permanent contract because this has happened for many other employees with fixed-term contracts in that company. Such implicit expectations often will only become apparent when they are not fulfilled, in this example when the employment contract is not prolonged. Breaches in psychological contracts are difficult to avoid because of their inherent opaqueness. At the same time, they may weigh as heavy or even heavier than breaches in formal contracts due to the loss of trust in the norm of reciprocity.

6.3 Past and Current Social Contracts for Risk and Safety Management

By considering the basic mechanisms of social exchange and including implicit expectations as captured by the notion of psychological contracts, a more fine-grained picture of social contracts for risk and safety management emerges. In the following, the social contract of the past and of today are discussed, illustrating how explicit and implicit expectations and obligations have changed in ways that have led to severe breaches in the contract.

For the longest time, business operations that imply risks for health, safety, and the environment were conducted with little or no regulation, resulting in human illnesses and deaths as well as environmental damages over centuries. Economic activities have come under stronger scrutiny only in the twentieth century, especially due to increased worker representation in unions and the rise of tripartite institutions such as the International Labor Organization. Moreover, the widespread strengthening of democratic values demanded controls against overly powerful social actors, such as protection of the public against irresponsible business practices.

These developments facilitated a social contract between the public, independent oversight bodies, and firms that promised the public to receive valued goods, such as electrical power, fuel, or rail travel with (close to) no risks. The main risks were perceived to be of a technical nature and were mitigated by engineers employed in the firms and oversight bodies who shared the fundamental belief that the impact of technology can be fully controlled. The public trusted this control, having no means of control themselves. This contract contained formal agreements, for instance regarding oversight by regulators, but also many implicit expectations creating a psychological contract built on technocratic assumptions.

Recurring accidents, continuing irresponsible business practices, and a general loss of confidence in the natural sciences as the main driver for societal progress have contributed to severe rifts in this social contract. Attempts at establishing a new social contract have led to arrangements that lack the fundamental glue of reciprocity. With the growing acknowledgment that technologies cannot be fully controlled due to their inherent complexity but also due to the complex interplay of technological and social factors in determining outcomes and impact, firms have been allowed to establish their own forms of risk management under the new regime of self-governance (May 2007). Regulators aim to set and monitor guardrails for this self-governance, but often lack resources, competence, and authority to do so effectively. The public reacts with an increasing sense of unease and mistrust, both concerning adequate risk management of business operations and the desirability of goods produced, for instance in relation to climate change, but lacks power and control to affect change. On top of it all, an emerging new world order, with countries at the forefront of technological innovation that lack or are starting to lose fundamental democratic values, appears to thrive on unvetted competition to secure economic success, with doubtful consequences for social welfare.

6.4 Hopes for a New Social Contract for Safety

In many contexts, such as trade and labor or the digital transformation, there are calls for a new social contract that reestablishes reciprocity of rights and obligations among different social actors (Triangle 2023). However, power imbalances among actors are enormous, making any fair deal unlikely. In the context of risks posed by artificial intelligence (AI), it has been argued that more powerful actors may be persuaded to agree on reciprocal social exchange because the complexity of AI can only be mastered by joint efforts aimed at aligning control over AI systems and accountability for their outcomes (Grote et al. 2024). Such alignment needs to be continuously negotiated during technology development and use because opportunities for control by different actors in the AI supply chain—such as developers, users, and providers of training data, computing power, or foundational models—change during the AI life cycle.

Might similar principles be applied in a new social contract for managing high-risk business operations more generally? I would argue yes and will develop that argument more fully based on three principles: control–accountability alignment, stakeholder interdependence, and integrative stakeholder negotiations. Examples from of AI development and use specifically in high-risk industries are used to support my reasoning. As will be seen, the suggested principles should allow for the development of social contracts that are adapted to the particular conditions of an industry, a world region, or a society because they are agreed upon, upheld, and transformed in negotiations among all relevant parties. By fostering open communication about reciprocal expectations and obligations of all stakeholders, the psychological components of the social contract become more explicit which also furthers a more sustainable social exchange.

6.4.1 Control–Accountability Alignment Is Core to a New Social Contract for Safety

A basic requirement for safety is that operations can be sufficiently controlled to avoid harm. With every new wave of technological development, the belief is revived that full automation will be the way to achieve full control over the complex socio-technical systems that high-risk business operations require. However, to date with each such wave, that belief has been eventually chattered, requiring human operators of these systems to remain the ultimate guardian of safety. Most recently, human oversight has been established in the EU AI Act as a fundamental requirement for the safe operation of AI systems. However, with the drastically increasing complexity of AI and other emerging technologies, these systems become opaque and unpredictable for their users and may leave little room for human decision-making. Thereby, a social contract is created that is unjust from the start because human operators are held accountable for the functioning and outcomes of systems they have little or no control over. This problem has been long acknowledged as the "ironies of automations" (Bainbridge 1983). It is exacerbated with current technologies, especially when complex forms of machine learning are employed, because such systems become opaque even for technology developers, discussed as the black box problem.

Only if control and accountability can be aligned for all actors in a socio-technical system, a fair social contract is possible. Accountability motivates actors to act in accordance with superordinate social norms as they need to explain and justify their actions to members of societal institutions or the public at large. Control, on the other hand, enables actors to achieve desirable and to avoid undesirable outcomes (Grote et al. 2026). In highly complex systems, control and accountability may have to be shared among stakeholders, as no actor is in a position to fully understand and predict work processes and outcomes. Rather, all actors need to collaborate to gain sufficient insights into business operations and the risks they entail. If accountability is shared among stakeholders in line with their limited and distributed control, diffusion of responsibility may result, where everyone assumes that someone else is responsible. To achieve the opposite, that is everyone is aware of the need to jointly accept accountability, accountability must become a shared norm rooted in the culture of all organizations and institutions involved. Oversight bodies have an important role to play as they may grant licenses to operate only if control–accountability alignment has been demonstrated for all actors involved.

Ongoing discussions at EASA, the European Union Aviation Safety Agency, illustrate the importance of control-alignment, but also the difficulties in achieving it. Trade union representatives resist attempts to aim for full automation even for operations where human oversight is essentially impossible due to the complexity of the respective technical systems. Fearing job loss, they seem to prefer being held to account for operations they cannot control anymore over explicitly giving up both control and accountability for those operations. This example shows that social contracts for safety need to be embedded in broader social contracts that establish obligations of economic and political actors to protect the livelihood and wellbeing of all members of their societies. Moreover, it demonstrates the importance of considering different layers of expectations and obligations in social contracts. Discussions on control–accountability alignment may apparently deal with factual requirements for all actors to keep a specific socio-technical system safe, while implicitly actors worry about losing stakes on other issues, such as their employment.

6.4.2 Stakeholder Interdependence Requires Decentralized Governance of Social Contracts for Safety

Control–accountability alignment can additionally be aided by the realization that in highly complex systems, stakeholders cannot act independently anymore. This is particularly apparent when AI is involved because system development is not possible without high-quality training data that must be provided by prospective system users or by third parties, while system use itself also shapes the data the systems learn from. However, also in other complex socio-technical systems, learning from successes and failures during pilot testing and routine use is crucial for mitigating system risks. Once stakeholders acknowledge these interdependencies, they will also find that traditional centralized forms of governance do not work anymore, as no one can claim sufficient knowledge and means of influence to be the dominant voice in risk management decisions (Bridoux and Stoelhorst 2022). More equalitarian and decentralized interactions among stakeholders are better suited to share knowledge and cope with the uncertainties inherent to running complex and possibly harmful operations. If such forms of governance can be established, it also becomes more likely that the social contract is built on true reciprocity.

Such a shift in thinking is not easy, however. Normally, actors in business relationships try to protect their independence and seek insurance against negative impacts of dependence, for instance concerning failed or faulty deliveries in supply chains. In an ongoing industry project in which I have been involved, major difficulties have arisen in the commissioning of an AI-based system for automatic detection of flaws in safety-critical components, because, among other things, both the system provider and the commissioning organization want to obtain full control over data used for training the system and data obtained during the use of the system. Negotiations with one provider have already failed, and the negotiations with another provider are currently stuck for similar reasons. Efforts are underway to help all parties involved to accept inevitable interdependencies and to become more open toward discussing different scenarios for sharing control and accountability.

6.4.3 *Integrative Stakeholder Negotiations Allow for Tailored Social Contracts for Safety*

The above example highlights the necessity to develop an understanding for the interdependencies in complex socio-technical systems to be able to establish settings in which stakeholders can successfully achieve control–accountability alignment for all. The aim is to foster integrative negotiations among stakeholders that are based on a mindset focused on fostering (smaller) gains for everyone rather than a winner takes it all attitude (Brett and Thompson 2016). To achieve an equitable social contract, it is key to be transparent about possible benefits and losses for different stakeholders and about different expectations and needs among the stakeholders. If such an open dialogue can be established, it also becomes easier to reveal implicit expectations and assumptions and thereby expose the psychological part of the social contract.

Negotiations about control–accountability alignment and further aspects of risk and safety management need to consider cultural, societal, and political circumstances to promote fair social contracts. In particular, power differentials between stakeholders have to be taken into account and measures taken to foster equal participation of weaker stakeholders. If a shared norm of accountability can be established, drawing on regulatory regimes that adequately balance constraints and empowerment, chances are increased that more powerful stakeholders accept accountability in line with their means of control, and the rights of less powerful stakeholders, such as workers especially in low-income countries, are upheld.

Research by the ILO in the context of AI supply chains, highlighting the invisible work of data annotators and content moderators in African and Asian countries, showcases the importance of widening the system boundaries for building a new social contract for safety. Discussions on the AI supply chain have helped to foster an understanding of the many highly interdependent actors from around the globe that are responsible for creating, running, and maintaining safe socio-technical systems. Most high-risk operations involve such large networks of actors with very different stakes and opportunities for becoming part of equitable social contracts.

6.5 Conclusion

I have argued that the social contract for safety has been breached in several ways, especially regarding the promise of full risk control through technology and business interests being kept in check by safety concerns. To establish a new contract for safety, technocratic assumptions need to be overcome which may not only be difficult for technology developers and regulators, but also for the public as they need to accept a new order of risk management that requires more informed decisions on how risks are to be mitigated. For workers in high-risk systems, the social contract has never been equitable as they usually are held to account for outcomes they do not have sufficient control over. This situation is exacerbated when those workers are employed in precarious settings that lack adequate worker protection. More recently, the social contract for safety has gotten under even more pressure, as the new geopolitical realities place a premium on unvetted innovation, thereby calling any regulatory action into question.

The three proposed principles—control–accountability alignment, decentralized stakeholder governance, integrative negotiations—can help to establish a new contract for safety but require a strong alliance among international institutions, such as regulatory bodies from various industry sectors and the ILO, to push for action. Having a broad international outlook on the path forward will foster consideration of the diversity of working and living conditions across the globe, hopefully preventing a further separation between the haves and have nots in this world. Easy it is not, given that not even regulatory agencies within one country seem to talk much to each other at this point.

References

L. Bainbridge, Ironies of automation. Automatica **19**, 775–779 (1983)
J. Brett, L. Thompson, Negotiation. Organ. Behav. Hum. Decis. Process. **136**, 68–79 (2016)
F. Bridoux, J.W. Stoelhorst, Stakeholder governance: solving the collective action problems in joint value creation. Acad. Manag. Rev. **47**, 214–236 (2022)
Encyclopedia Britannica. https://www.britannica.com/topic/social-contract (2025)
A.W. Gouldner, The norm of reciprocity: a preliminary statement. Am. Sociol. Rev. **25**, 161–178 (1960)
G. Grote, S.K. Parker, K. Crowston, Taming Artificial Intelligence: a theory of control–accountability alignment among AI developers and users. Acad. Manag. Rev. **51**(2), 278–299 (2026)
E.A. Lind, T.R. Tyler, *The Social Psychology of Procedural Justice* (Plenum Press, New York, 1988)
P.J. May, Regulatory regimes and accountability. Regul. Gov. **1**, 8–26 (2007)
D. Rousseau, Psychological and implied contracts in organizations. Empl. Responsib. Rights J. **2**, 121–139 (1989)
L. Triangle, Why a new social contract is urgent? Int. J. Labour Res. **12**, 6–18 (2023)

Chapter 7
Risk Regulation at a Crossroads or Dead End? Ideas for the Road Ahead for Innovative and Sustainable Regulatory Approaches

Siri Wiig and Catherine Calderwood

Abstract This chapter investigates current challenges in risk regulation, raises ideas, and suggests how regulators may need to innovate, collaborate, and rethink their approaches for handling risk with available resources in sustainable ways. We use examples from healthcare and petroleum industry to illustrate how regulators (1) need to deal with highly competent people and organizations where risk develops in new ways and (2) tackle systemic challenges in modern society.

Keywords Risk regulation · Strategy · Approach · Innovation · Sustainability

7.1 Introduction

It is often stated that risk regulation lags behind practice and how new risks develop in any sector (Wiig et al. 2024a, 2024b; Kirwan et al. 2002). Technological innovations, changing work processes, procedures, and practices, fast changing society, ever-increasing specialization, and diverse threats and emerging risks all challenge how regulators can regulate, and approach regulated bodies within their jurisdiction. To add further complexity to this, the fact that safety is a moving target (Vincent and Amalberti, 2015) implies that regulators continuously need to evaluate, innovate, and test new strategies and approaches to keep up with the pace.

Modern society with highly knowledgeable and informed consumers and organizations also brings new expectations, opportunities, and challenges for regulators in different sectors. To exemplify, we see how modern workers are aware of their rights

S. Wiig (✉)
Faculty of Health Sciences, SHARE Centre of Resilience in Healthcare, University of Stavanger, Stavanger, Norway
e-mail: siri.wiig@uis.no

C. Calderwood
University of Strathclyde, Glasgow, Scotland, UK
e-mail: catherine.calderwood@strath.ac.uk

C. Bieder et al. (eds.), *Living with the New Safety Landscape I*,
SpringerBriefs in Safety Management, https://doi.org/10.1007/978-3-032-29336-7_7

in high-risk industry; leaders hold a great responsibility across sectors of follow up many types of legislation related to different risks (occupational safety, major accident risk, security) (Magerøy et al. 2024; Oldenhof and Bal 2023); consumer power and influence must be taken into account; patients and families are highly knowledgeable about their illness and rights; and across sectors we see an increased need for documentation and transparency to maintain trust in services and professionals (Han 2015; Wiig et al. 2018). A systemic challenge in this field furthermore relates to how regulated organizations change their approach risk, adherence to regulation, and instead of complying to regulations, sometimes prefer to go into power battles with regulatory bodies for example when large financial interests may be involved or when innovation is hampered by current regulation.

Do we ever think about this complexity as a potential problem, or trap that society has put itself into? A society where people, leaders, workers, organizations, and regulatory officials all perhaps face unrealistic, high expectations resulting in cover up, defensive medicine or work practice, and potentially resulting in counterproductive use of limited recourses? One may question if modern risk regulation is at a crossroads or a potential dead end as more regulation, stronger follow-up, the use of rewards or sanctions toward individuals and organizations, all may cause trouble for both the regulators and regulated. This is because it would be impossible to develop detailed regulation covering all types of risks in complex adaptive systems, and it would be impossible to inspect and enforce regulation and a sanctioning regime in fair ways. So, how can this be handled, what are the future scenarios we may face and need to consider, and what could potentially be suggested as innovative and sustainable approaches in risk regulation in more and more complex adaptive systems?

This chapter aims to reflect on these questions, and addresses some of the emerging topics within risk regulation that challenge the current regulatory logic of compliance, evidence, and risk signals in light of what modern accident theory sees as key for understanding and managing risk and safety.

7.2 Trends and Challenges in Risk Regulation

The traditional command and control approach to risk regulation put high responsibility on the regulators to prescribe regulation, inspect and to sanction regulated bodies such as businesses and public institutions. The regulated bodies on the other hand need to comply to specific and detailed rules and a distrust characterizes the relation between regulator and regulated. The approach is challenging as regulators need to know details about risk to prescribe specific regulation, and it is considered to be at one end of a continuum (Refslund 2022; Wiig and Tharaldsen 2012).

On the other end of the continuum, one will find approaches such as enforced self-regulation, self-regulation, and what is referred to as responsive and reflexive regulation (Ayres and Braithwaite 1992; Baldwin and Cave 2012; Refslund 2022; Øyri et al. 2024). The latter approaches are focusing on a more goal-based or performance-based regime where the regulators set goals for the regulated practice and leave the solutions with the regulated organizations. This approach relies more strongly on trust between the regulators and the regulated and is also seen as a better way to enable adaptive capacity (Øyri et al. 2024).

However, there are challenges in succeeding in goal-based approaches, and practices vary in companies. Refslund (2022) found that degree of compliance with goal-based regulation was largely depending on internal company conditions, especially worker involvement, professionalization, and management support. If these conditions were not present, it resulted in non-compliance which challenges the policy makers and the means of dealing with regulatory practice and sanctions. When your regulatory approach is anchored in trust, involvement, and reflexivity and this fails, what would your response be?

7.3 Practical Examples

There exist cross sector challenges and varying regulatory regimes, but they all face the need to innovate and improve regulatory practice. In the following paragraphs, we would like to exemplify why regulators are challenged by societal demands, companies, institutions and workers, consumers and patients and dynamics in these regulatory environments. We will illustrate with different examples, where we first show power issues between regulator and regulated organization; second we illustrate distribution of responsibility within socio-technical systems; in the third we break down a case at the individual level and how regulators need to address cases of systemic issues manifesting at the individual level. These examples are chosen based on research projects, current literature, and authors background from societal safety, petroleum regulation, and as practicing doctor.

First, we will illustrate with an example from oil and gas industry where what was seen as too much dialogue between the regulator and the regulated resulted in requests for stronger control measures and more use of sanctions. The Norwegian oil and gas regulatory regime has for years practiced enforced self-regulation with strong reliance on dialogue, systems perspective, tripartite collaboration between the regulator, the companies, and unions. The regime is characterized by a dialogue-based approach and limited use of strong sanctions, even though the regulation allows for shutting down a company and recommending police investigation and prosecution. Over the years, this approach has worked, but we have seen requests from the unions and Norwegian Audit Office to use stronger sanctions as the large companies have been accused of not complying with the regulation with the current use of measures. Compared to, e.g., the Norwegian Labour Inspection Authority (NLIA), the Norwegian Ocean Industry Authority (NOIA) does not report to the police or fine companies, while the NLIA does several times a year. This shows a major difference within the same country. Companies use their power and their position to resist recommendations, protest, and making regulatory practice more challenging and ineffective. The "cozy approach" focusing on dialogue has its limits some say, and the NOIA has changed practice and uses more sanctions currently, compared to some years ago (Aftenbladet 2024; Riksrevisjonen 2019). However, the power dynamics are continuously challenging regulators, and their approaches are challenged by companies from different countries, sizes, and strategic aims.

Secondly, we will use an example showing how risks can attenuate or amplify through the socio-technical system as different actors have different understanding of what risk is and how it is handled (Calderwood et al. 2024; Idsøe-Jakobsen et al. 2025; Wiig 2012). This reality challenges regulatory practice as systems become more fragmented, and risks may be hidden for regulators, although workers argue it is increasing, changes in risk level is sometimes hard to document.

In healthcare we see a current crisis in the lack of healthcare personnel. This crisis is predicted to grow, and the system is already a stretched system. Homecare services, which are a complex adaptive socio-technical system, are considered to be a key part of the solution (Idsøe-Jakobsen et al. 2025). This implies that patients will be treated in their homes for increasingly complex conditions to a much larger extent. This requires higher competence levels from the workforce, more advanced technology and increased availability of medications and an assessment and acknowledgment of the diverse risks that may be in the patients' homes. Policy makers advocate for different homecare solutions and use of telecare and monitoring, without necessarily assessing potential risks that come with these changes. Policy makers thereby transfer the responsibility of how to operationalize service provision to complex patients, to the municipalities, in a Norwegian setting. The municipalities are, according to the law, responsible for ensuring safe services to service users. Policy makers seem to attenuate risks, and the homecare service providers are the ones facing real-life risks but often without enough resources (funding, personnel, competence). When something goes wrong and patients are harmed, healthcare regulators tend to look closely at the active failure and not at the systemic problems resulting in working conditions out of line with requirements for performing sound professional practice. Regulators also need to take a stronger interest in understanding and approaching risk when system actors are attenuating risks downwards through the system (Idsøe-Jakobsen et al. 2025). This is not in line with the current logic of evidence where regulators search for "evidence" to document is rules were broken or not, in linear ways. In complex adaptive systems, understanding of causality is not linear. New ideas need to be developed for regulatory innovation to prevent blind spots, normalization of deviance, and better understanding of how risk emerges and can be prevented.

Thirdly, we will share an example of regulatory challenges which can be related to critical infrastructure break down and how this ends up as a potential individual problem. When technological systems are suddenly not working in an organization that cannot stop its operations, this will heavily rely on individual workers' competence and capacity. Who will be responsible if something goes wrong, and what is needed to succeed? A real example comes from one of the authors (CC) who works as a consultant obstetrician providing emergency care to pregnant women who are referred or may call and present themselves in labor or with a variety of significant antenatal and past partum problems. The unit sees 155 patients per month and is open 24/7 staffed by experienced midwives and doctors. On the day of new doctor changeover which happens for all trainee doctors once per year the hospital had planned an upgrade of all the electronic patient records including all clinical notes, past and present, laboratory orders and results, referrals, prescribing and all previous patient information and correspondence including the ability to make contemporaneous notes.

Unfortunately, the planned upgrade caused the entire electronic system to fail. None of the described functions was available for 48 h. CC was the only senior doctor in this emergency department, training the new doctors, many of whom had not been in a maternity unit previously. Other colleagues were in clinics and on labor ward, all without any information except what was being given by the patients. There were no blood results or other diagnostic test results available and ordering new tests had to be done manually. As one can imagine this led to a very unsafe and unstable situation, risking incorrect or inadequate treatment and with no ability to record except on paper for which there is no longer an adequate filing system. CC is not aware of any adverse outcomes although there were several examples of "near miss" when careful history taking averted major issues. What if there had been an adverse event leading to an adverse outcome? The incorrect treatment being given or worse still a previous significant condition not being taken into account? Who would have been responsible or to blame and where would the regulator for this doctor (in this case the General Medical Council) find fault? Should that individual doctor be held to account and face sanctions to their practice?

This example demonstrates how complex current patient treatment is, and how challenging it would potentially be for a regulatory body to ensure compliance to the regulation and follow-up of potentially both individual professionals and the organization and how regulators need to balance between diverse legislations and potential use of sanctioning means (Wiig et al. 2024a, 2024b).

7.4 Possible Ways for the Road Ahead

In this chapter, we have illustrated some of the current challenges in risk regulation. We argue future regulatory practice needs to look beyond active failures and promote a system perspective, and better understanding how too much or too little trust can result in an unbalance in the regulatory dynamic. Here we will suggest some possible solutions and what can be considered as ways forward for systemic improvement:

Acknowledge High-Risk Systems Are Too Complex to Blame Individuals. The only acceptable way for regulators to looking at individuals is in situations of sex, drugs, and rock'n'roll as it is referred to in healthcare, or when workers and leaders intentionally fail in any sectors (Wiig et al. 2024a, 2024b). Accountability will always be important, but in large complex socio-technical systems with increasing tasks, lack of personnel there must be caution to only look at the individual worker when something goes wrong.

Seek Regulatory Innovation Driven from within. This requires interest among regulatory bodies to innovate, test, evaluate, and take part in research to ensure evidence-based regulation and practice relevant regulation. It requires funding and sandboxes and dialogue (Macrae and Ansell 2024) and willingness to change, which has sometimes proved hard for some regulators (Wiig et al. 2018).

Emphasize the Role of Soft Signals, Culture, Leadership and to Learn from What Goes Well. One may wonder what role regulators can have in this logic—challenging the fundamentals of objective evidence in risk regulation, control, and law enforcement (Wiig et al. 2024a, 2024b). Regulators can innovate and pay more attention to these soft elements as fundamentals of risk management, with anchoring in literature (Kok et al. 2020). Regulators could also seek to develop better competence and support leadership programs for the regulated organization, for example the EU railways regulatory body which runs targeted programs for leaders to understand the role of human factors, leadership, and the importance of good relationships in their work practice.

Accept Workers' Articulation of Workarounds. We know risk will emerge, accidents happen, and we know we can learn from what goes well. Regulators need to acknowledge the importance of adaptive capacity among individuals and regulated organization. Regulators need to allow the practice field to articulate their workarounds to either make sure that irrelevant procedures and outdated regulatory frameworks are removed, or that new well-functioning work practices are integrated into procedures. Regulatory bodies need to appreciate front-line competence, autonomy, and work—when they succeed and when they don't without punishing.

Collaboration across sectors and use of advanced AI tools and analytics to provide up to date information that aids understanding and handling the complexity in high-risk industry and how sectors are connected in modern crises and normal operation. Regulators may perhaps play a stronger role in the future to support the regulated bodies with information, tools, methods to help them learn new ways of understanding safety and quality in when living with crises.

7.5 Conclusion

In this chapter, we have seen the complexity of regulating and being regulated in modern society. Future regulatory approaches need to be adaptable to new types of risks, new types of organizations, and new ways of using information. Innovative regulatory capacity and adaptive capacity in organizations are both fundamental for the road ahead. This will have consequences for resources, power, and competence. Regulatory bodies, by construction, have financial, institutional, and legal constraints. It's not enough that they want to learn and adapt—the entire regime needs to be set up and equipped to understand and handle emerging risks in complex adaptive systems from the shop floor to the regulators as future risks will most likely develop in ways we cannot predict and meet with old approaches.

References

Aftenbladet, (2024). Havindustritilsynet har aldri politianmeldt et oljeselskap. Arbeidstilsynet politianmelder 50 bedrifter i året. (26.4.2024)

I. Ayres, J. Braithwaite, *Responsive Regulation: Transcending the Deregulation Debate* (Oxford University Press, 1992)

R. Baldwin, M. Cave, M. Lodge, *Understanding Regulation: Theory, Strategy, and Practice* (Oxford University Press, 2012)

C. Calderwood, G.S. Braut, S. Wiig, What is truly informed consent in medical practice and what has the perception of risk got to do with it? Healthcare **2025**(13), 8 (2024). https://doi.org/10.3390/healthcare13010008

I. Idsøe-Jakobsen, H. Dombestein, R. Lawton, S. Wiig, Adaptive capacity within the homecare setting: the importance and value of a shared perception of risk. J Patient Saf. (2025), published ahead of print. https://doi.org/10.1097/PTS.0000000000001323

B. Kirwan, A. Hale, A. Hopkins (eds.), *Changing Regulation, Controlling Risks in Society* (Pergamon, Oxford, 2002)

J. Kok, I. Wallenburg, I. Leistikow. R. Bal, The doctor was rude, the toilets are dirty. Utilizing 'soft signals' in the regulation of patient safety. Saf. Sci. **131**, 104914 (2020). https://doi.org/10.1016/j.ssci.2020.104914

L. Oldenhof, R. Bal, Regulatory leadership: Conducting mundane work to 'tailor' rules, in *Research Handbook on Leadership in Healthcare*, ed. by N. Chambers, (Edward Elgar Publishing, 2023), pp. 388–401. https://doi.org/10.4337/9781800886254.00030

C. Macrae, C.K. Ansell, Generative spaces: collaboration, learning and innovation in a regulatory sandbox (May 13, 2024). Available at SSRN: https://ssrn.com/abstract=4825907 or http://dx.doi.org/10.2139/ssrn.4825907

M. Magerøy, C. Macrae, G.S. Braut, J. Braithwaite, R. Clay-Williams, S. Wiig, Leading quality and safety on the frontline—A case study of department leaders in nursing homes. Journal of Healthcare Leadership **2024**(16), 193–208 (2024) https://www.dovepress.com/getfile.php?fileID=98545

Riksrevisjonen, Riksrevisjonens undersøkelse av Petroleumstilsynets oppfølging av helse, miljø og sikkerhet i petroleumsvirksomheten. Dokument **3**, 6 (2019) (2018–2019). https://www.riksrevisjonen.no/globalassets/rapporter/no-2018-2019/petroleumstilsynet.pdf

C. Vincent, R. Amalberti, Safety in healthcare is a moving target. BMJ Qual. Saf. **2015**(24), 539–540 (2015)

B. Refslund, How self-regulating can it be? Explaining limitations in firms' compliance with reflexive regulation. Int. Rev. Public Policy [Online], 4:1 2022, Online since 01 April 2022, connection on 28 January 2023, URL: http://journals.openedition.org/irpp/2590; doi: https://doi.org/10.4000/irpp.2590

S. Wiig, C. Calderwood, B. Høie, J. Braithwaite, Sex, drugs and rock'n roll: The only reasons for regulators to target individuals. BMJ Qual. Saf. (in press) (2024a). https://doi.org/10.1136/bmjqs-2024-017214

S. Wiig, H. Lyng, J. Braithwaite, D. Greenfield, C. Calderwood, Foundations of safety—realistic medicine, trust, and respect between professionals and patients. Int. J. Qual. Health Care **36**(1) (2024b). https://doi.org/10.1093/intqhc/mzae006

S. Wiig, K. Aase, M. Bourrier, O. Røise, Transparency in health care—displaying adverse events to the public, in *Risk Communication for the Future*, ed. by M. Bourrier, C. Bieder (SpringerBriefs in Safety Management, 2018), doi:https://doi.org/10.1007/978-3-319-74098-0_8, pp.111–125

S. Wiig, J.E. Tharaldsen, In regulation we trust. Work.: J. Prev., Assess. Rehabil. Suppl. **41**(1), 3043–3050 (2012) Available at: http://iospress.metapress.com/content/p76u7168057g6214/?p=9219b522da3f419fb735dfe7f1593a0e&pi=0
S.F. Øyri, J. Braithwaite, D. Greenfield, S. Wiig, Resilience and regulation—Antithesis or a smart combination for future healthcare service improvement. Int. J. Qual. Health Care **36**(1) (2024). https://doi.org/10.1093/intqhc/mzae002
B.-C. Han, *The Transparency Society* (Stanford University Press, Stanford, 2015)

Chapter 8
Health and Safety in Precarious Work: Eroding Foundations of the Nordic Model of Labor Regulation

Ulla Forseth and Hilde Bjørkhaug

Abstract This chapter examines how the growth of precarious work challenges established systems of occupational health and safety, regulatory institutions, and workers' health, safety and wellbeing within the Nordic model of labor regulation. While the Nordic countries are often portrayed as exemplars of regulated employment relations and strong worker protection, we demonstrate how increasing forms of precarious employment in Norway expose significant gaps and blind spots. Drawing on empirical examples from sectors prone to precarization, agriculture, construction, and civil aviation, we illustrate how both unregulated workers and workers formally covered by existing regulations experience deteriorating working conditions with potential negative consequences for health and safety. By examining these developments, the chapter highlights key structural tensions inherent in capitalist economies and poses a fundamental question for labor governance: can existing labor policies and regulatory frameworks accommodate these transformations, or is substantial revision required to ensure future health and safety at work?

Keywords Precarization · Health and safety · Regulation · Nordic model · Structural changes

U. Forseth (✉) · H. Bjørkhaug
Department of Sociology and Political Science, Norwegian University of Science and Technology, Trondheim, Norway
e-mail: ulla.forseth@ntnu.no

H. Bjørkhaug
e-mail: hilde.bjorkhaug@ntnu.no

C. Bieder et al. (eds.), *Living with the New Safety Landscape I*,
SpringerBriefs in Safety Management, https://doi.org/10.1007/978-3-032-29336-7_8

8.1 Introduction

Dominant models of occupational safety and health are largely built on assumptions of stability and predictability: that risks can be identified in advance, managed through standardized procedures, and controlled via audits, inspections, and compliance mechanisms. These models also rest on the idea that safety can be effectively governed through dedicated institutional arrangements, such as regulatory authorities, organizational safety departments, and collaboration between the state, employers, and labor unions or professional associations. In a globalized labor market characterized by rapid change, atypical employment relations, insecurity, and instability, these assumptions are increasingly challenged. Precarious employment arrangements can directly affect workers' health, safety, and wellbeing by weakening job security, reducing access to training and representation, and limiting workers' capacity to voice concerns or refuse unsafe work. Processes such as digitalization, de-unionization, labor migration, and the shifting of regulatory power from public authorities to private actors have contributed to the erosion of traditional employment relations and established forms of worker protection.

The Nordic model is an ambiguous concept (Byrkjeflot et al. 2022) but is widely recognized for its universal welfare systems, comprehensive regulation and emphasis on dialogue and bi- and tripartite collaboration between employers, labor unions and the state in promoting decent work, health and safety (Bungum et al. 2015; Alsos et al. 2024). In high-risk industries such as oil and gas, such collaboration has been identified as a key factor underpinning high safety standards (Rosness and Forseth 2014). Although the Nordic model has often been described as resilient in times of crisis, an increasing number of cases suggest that it struggles to address new forms and processes of precarization that fall outside, or sit uneasily within, existing regulatory frameworks.

These challenges are further compounded by a continued reliance on established regulations, protocols, and monitoring systems that assume stable employment relationships and clearly identifiable employers. As work becomes more fragmented, transnational, and flexible, such frameworks may be ill-equipped to ensure effective protection of workers' health and safety. Against this backdrop, the chapter addresses the following research question: *How do precarious employment and processes of precarization challenge workers' rights and protection within the Nordic model of labor regulation?*

Following this introduction, we clarify key concepts related to precarious work, precarious employment patterns, and precarization as an ongoing process. We then present empirical examples from strategically selected industries with high levels of precarious employment, as well as from a global industry where processes of precarization have reshaped working conditions, health, and wellbeing despite formal regulatory coverage. The chapter concludes with a discussion of the implications of these findings for labor regulation and occupational health and safety in the Norwegian context and reflects on whether the Nordic model can adapt to or must fundamentally rethink its approach to an increasingly precarious future of work.

This chapter makes three concise contributions to research and practice on occupational health and safety within the Nordic model of labor regulation. First, it advances existing research on precarious work by explicitly linking processes of precarization to the functioning and limits of dominant OHS models and regulatory frameworks. Rather than treating precarious work solely as a matter of employment conditions or labor market inequality, we show how precarization reshapes risk exposure, safety management practices, and workers' ability to exercise rights related to health and safety. Second, the chapter contributes empirically by analyzing precarization across both traditionally weakly regulated sectors and highly regulated, safety-critical industries. By juxtaposing agriculture and construction with civil aviation, we demonstrate that precarization is not confined to regulatory "grey zones," but can also undermine health, safety, and wellbeing within sectors that formally comply with Nordic labor regulations. This comparative approach reveals how regulatory coverage alone is insufficient to guarantee effective protection under conditions of fragmented and flexible employment. Third, the chapter contributes to debates on the future of the Nordic model by identifying structural tensions between established forms of labor governance and emerging employment relations. We argue that current regulatory instruments, monitoring practices, and collaboration mechanisms are increasingly misaligned with the realities of precarious work. In doing so, the chapter provides analytically grounded insights for policymakers, regulators, labor unions, and employers seeking to maintain high standards of health and safety in an evolving and more precarious world of work.

8.2 Precarious Employment, Precarization Processes, Health, Safety, and Wellbeing

The ambition and reputation of the Nordic model rest on ensuring that all stakeholders in the labor market have the right to participate in negotiations concerning working conditions. The overarching aim is to promote secure employment, safe working environments, and opportunities for wellbeing and personal development. In principle, this framework applies to workers who are covered by labor laws and regulations.

The terms *precarious work* and *precarious employment* are increasingly used to describe evolving labor market patterns and emerging employment trends where workers have little agency and low protection at work. Growing internationalization and increased labor mobility have contributed to the emergence of sectors and employment arrangements where the presence of precarious work and processes of precarization must be critically examined, even within the Nordic and the Norwegian labor market.

Kalleberg (2018) identifies two main theoretical perspectives that have shaped research on precarious work. The first focuses on new contexts for organizing "old" forms of work and on the transfer of risk from employers and the state to workers.

Kalleberg and Vallas (2018:1) define precarious work as "work that is uncertain, unstable, and insecure and in which employees bear the risks of work (as opposed to businesses or the government) and receive limited social benefits and statutory protections." This definition aligns with that of the European Agency for Safety and Health at Work (2024). Precarious work thus falls below socially accepted standards in which workers are entitled to certain rights and employment protections associated with economic participation (Kalleberg 2018, p. 13). It represents a strategy through which employers shift risks and responsibilities onto workers, rendering their employment conditions precarious.

A systematic review by Koranyi et al. (2018) of the relationship between precarious employment and occupational accidents and injuries concluded that certain dimensions of precarious employment increase the risk of workplace injuries. In 10 of 17 studies, the authors found a positive association between precarious employment and occupational injuries, particularly among individuals holding multiple jobs and those employed by temporary agencies or subcontractors at the same worksite. However, they emphasize the need for more high-quality, prospective studies to establish causality and effect sizes. Lau et al. (2024) further show that migrant workers face a significantly higher risk of fatal occupational injuries, a disparity attributed to structural factors such as precarious employment arrangements and inadequate safety protections. They argue that labor protection legislation must explicitly include migrant workers and that working conditions must be substantially improved to protect this growing segment of the workforce.

The second perspective outlined by Kalleberg (2018) adopts a broader structural lens, focusing on transformations within contemporary capitalism that enable the exploitation of labor. While precarious employment is not a new phenomenon, processes such as globalization, financial volatility, and the transnational mobility of labor and capital have intensified its prevalence, particularly in affluent democracies. These developments have far-reaching implications for job and income security, family stability, and overall wellbeing (Kalleberg 2018, p. 9). From this perspective, precarious work should be understood not merely as an individual experience, but as a structural process unfolding over time.

Alberti et al. (2018) further develop this approach through the concept of *precarization as a process*, emphasizing its dynamic and multifaceted character. The concept allows for a more nuanced analysis of the drivers, patterns, and manifestations of precarious work, including both explicit mechanisms such as contractual differentiation and more implicit ones, such as managerial practices that cultivate insecurity among workers (Alberti et al. 2018, p. 451). This processual perspective highlights how precarization evolves over time and captures the lived experiences of workers navigating insecure conditions. Importantly, it draws attention to the emergence of precarization within traditionally stable and prestigious occupations, challenging the assumption that precarious work is confined to marginal or low-status jobs.

The perspectives on precarious employment and precarization presented above offer a lens for analyzing how insecurity and risk are produced in contemporary labor markets. In the sections that follow, this framework is applied to three sectors;

agriculture, construction, and civil aviation, to show how precarization processes take shape in different institutional and organizational contexts, and with distinct implications for health, safety, and wellbeing.

8.3 Sector Studies

We present three sectors that illustrate distinct, yet interconnected challenges related to occupational health and safety, as well as processes contributing to increasing precarization.

8.3.1 Agriculture

According to estimates by the International Labour Organization (ILO 2020), agriculture employed approximately 874 million workers globally in 2020, accounting for 27.4% of the global workforce. Despite being a major employer, agriculture is also responsible for an estimated 210,000 work-related deaths annually. Agricultural workers are three times more likely to die at work than workers in other sectors. While occupational fatalities are declining in most industries, the ILO reports an increase in agriculture.

This troubling pattern is also evident in Norwegian agriculture, albeit on a smaller scale. According to Statistics Norway (SSB 2024), agriculture accounted for 11 of 36 occupational fatalities in 2023, making it the sector with the highest number of fatal accidents. The sector also reported 6.7 occupational accidents per 1000 employees, compared to a national average of 3.9 across all industries.

Research on accidents and fatalities in Norwegian agriculture attributes these high rates to both structural and cultural factors. The sector has undergone profound transformation. In 2024, there were 36,790 agricultural holdings, a sharp decline from previous decades (SSB 2024). While three out of four farms have closed over one generation, remaining farms have expanded to maintain overall production levels. The traditional family farm has evolved into a professionalized and mechanized workplace, often operated by a single farmer who may also hold off-farm employment (Bjørkhaug and Blekesaune 2008). This shift has increased isolation, long working hours, and high workloads, all of which elevate risk exposure. Accidents are frequently linked to inadequate safety practices related to machinery and tractor use, driven by time pressure and insufficient training in complex, multi-risk environments involving animals, machinery, and advanced systems. Cultural norms valorizing the "hardworking, risk-taking farmer" may further encourage unsafe behavior, while stress and mental health challenges can increase risk tolerance (Ernstsen et al. 2025). Farmers are legally responsible for health, safety, and environmental (HSE) compliance for both themselves and their employees.

In parallel with the rise of single-operator farms, Norwegian agriculture has become increasingly dependent on hired labor, particularly in fruit and vegetable production (Rye et al. 2018). Migrant agricultural workers often experience precarious employment, poor working conditions, and significant health challenges (Urrego-Parra et al. 2022). The agri-food sector's reliance on a flexible, low-rights workforce frequently unorganized, unaware of labor protections, and socially disconnected from local communities raises concerns about precarity, sustainability, and equity (Constance et al. 2023).

The COVID-19 pandemic brought the situation of migrant agricultural workers into sharper focus and intensified existing socio-economic and health inequalities (Côté et al. 2021). Using the pandemic as a case, Bjørkhaug and Brobakk (2024) documented a "striking silence and lack of concern for the welfare of the individual migrant worker taking on extra risk during a global pandemic" (p. 86). Despite their essential role, migrant workers had limited agency in shaping their working conditions. The regulatory facilitation of high-risk migrant labor reveals precarization as an ongoing process and represents a departure from a model that traditionally positions workers as central actors in negotiations. The authors conclude that parts of Norwegian agriculture rely on precarious labor in ways that fall outside accepted norms of fairness, as defined by Kalleberg (2018).

8.3.2 Construction

In 2020, approximately 280.4 million people were employed globally in the construction industry, making it one of the largest employers worldwide (ILO n.d.). The sector plays a critical role in building infrastructure, housing, and other essential structures, yet it is also among the most hazardous industries globally. Construction workers are exposed to a wide range of risks, including falls, slips and trips, equipment-related accidents, electrical hazards, vibration and noise exposure, hazardous materials, confined spaces, and environmental stress. Employment in the sector is often unstable due to project-based work, and the use of temporary and seasonal labor, frequently recruited through staffing agencies, is widespread.

The preparation for the 2022 FIFA World Cup in Qatar represents an extreme example of the exploitation of migrant construction workers. Thousands of South Asian workers were reportedly exposed to hazardous and potentially fatal working conditions, extreme heat, inadequate training, and substandard living conditions, resulting in widespread loss of life. This case illustrates the darker sides of global capitalism, where human rights violations and worker exploitation occur in the pursuit of profit under conditions of weak regulation and insufficient enforcement (Forseth 2025). Historically, parallels can be drawn between such projects and monumental constructions built through coerced labor, characterized by high rates of injury and death.

The construction sector is also among the fastest-growing industries in Norway and one of the mainland industries with the highest number of fatal accidents

(Wasilkiewicz et al. 2018). Between 2012 and 2021, 79 construction workers died in occupational accidents. In 2023, the sector recorded 5.7 occupational accidents per 1000 employees, compared to 5.4 in industry (SSB 2024). Falls are the most common cause of accidents, followed by impacts from objects, cuts from sharp tools, and electrical shocks.

Following the 2004 enlargement of the European Union, Norway experienced an influx of migrant labor from Eastern Europe, much of it recruited through staffing agencies. Many migrant workers were employed in temporary positions characterized by low wages, poor housing standards, insecure employment, and elevated injury risk. Polish workers now constitute the largest migrant group in Norway and are well represented in construction (Wasilkiewicz et al. 2016). Coming from different work cultures, many are unfamiliar with the Nordic model's emphasis on dialogue and a combination of conflict and cooperation between unions and management (Forseth and Rosness 2021). Language barriers, limited knowledge of health and safety standards, and fear of dismissal may lead them to accept tasks that domestic workers would refuse.

8.3.3 Civil Aviation

Civil aviation was historically one of the most highly regulated industries, governed by international, bilateral, and national rules. Deregulation in the USA in the 1970s and in Europe in the 1980s, followed by the introduction of the EU's "Open Skies" policy, fundamentally transformed the industry (Forseth 2022). While deregulation increased competition, expanded routes, and reduced ticket prices, it also intensified pressure on wages and working conditions.

The situation of cabin crew is particularly illustrative. Research shows that what was once a relatively well-paid occupation in Norway, initially dominated by women and later increasingly by men, has become a low-income job (Forseth 2022). To keep their competitive edge, some airlines have increased working hours and the number of flight legs per day, pension schemes have been reduced, and other benefits have been cut. Intensified work schedules, fewer rest periods, and limited recovery time place considerable strain on health and work–life balance. Cabin crew in Norway report higher levels of health problems than many other occupational groups, and time pressure, stress, and fatigue appear to negatively affect safety performance (Ydersbond 2022).

This development can be understood as a process of precarization that also affects other occupational groups in the aviation sector and reshapes the role of trade unions. Whereas unions previously focused on improving employment conditions through collective bargaining, their efforts increasingly center on preserving existing rights and benefits. Traditional network carriers have adopted elements of low-cost airline business models, including the use of subsidiary companies employing lower-paid crew. As management seeks to maximize productivity in a competitive market, unions are forced into continuous monitoring and micro-negotiations. Although most cabin crew in Norway still hold permanent positions and are unionized, a once secure and well-paid occupation is being transformed into a low-paid, transitory job characterized by demanding working conditions.

8.4 Living with or Envisaging a Different Future?

Drawing on empirical examples from three sectors, this chapter has examined whether and how the Nordic model of labor regulation is challenged by emerging forms of precarious work. The Nordic model is traditionally characterized by strong labor rights, comprehensive regulation, and broad consensus around the role of trade unions in safeguarding employees' interests vis-à-vis employers and the state. Across the sectors analyzed, this institutional architecture remains formally intact. Umbrella organizations such as the Norwegian Confederation of Trade Unions (LO) and PARAT (independent trade union affiliated with the Norwegian Confederation of Vocational Unions (YS)) continue to play key roles in representing workers in construction and aviation, while agriculture participates in tripartite negotiations with the food industry and the state through organizations such as the Norwegian Farmers' Union.

In principle, such arrangements should limit the scope for precarious employment. This makes the presence and persistence of precarization within the Nordic model particularly significant. How, then, do precarious forms of work gain a foothold in a labor market designed to prevent them? And what do these developments imply for the future of working conditions, health, and safety?

8.4.1 Precarious Labor and Gaps in Protection

Our first observation is that precarious labor is not an external anomaly, but a growing feature within the Nordic labor market. Although this chapter has focused on agriculture, construction, and civil aviation, similar dynamics can be observed in other sectors, including cultural work and platform-mediated services (Nilsen et al. 2022). Precarious employment thus cuts across sectoral boundaries and challenges core assumptions of the Nordic model. Migrant workers are particularly visible in agriculture and construction, where they often occupy essential positions under conditions that diverge markedly from standard employment norms. These workers frequently face unstable contracts, limited access to union representation, and inferior living and working conditions. Our second observation is that existing protection mechanisms often fail to secure their rights in practice. While employers formally carry responsibility for health, safety, and working conditions, enforcement and accountability remain uneven. This allows the state, in effect, to delegate responsibility without ensuring that protections are upheld.

The COVID-19 pandemic made this tension particularly visible in agriculture. Here, the state primarily framed its role as ensuring access to labor to maintain domestic food production, while issues of worker protection received less attention. This illustrates a broader contradiction within the Nordic model: although designed to prevent precarious employment, certain political and structural choices may inadvertently enable it, especially in times of crisis.

8.4.2 Structural Cracks and the Limits of Inclusion

When groups of workers remain outside the collaborative institutions of the Nordic model, its capacity to deliver protection weakens. Trade unions often lack membership among migrant workers and therefore have limited ability to represent their interests. At the same time, migrant workers themselves are frequently constrained in exercising agency due to economic dependency, language barriers, limited contractual protection, and lack of knowledge about their rights. These dynamics are evident in both construction and agriculture, where employers have adapted to a labor force composed partly of migrant workers in precarious positions. Similar processes are now visible in civil aviation, where foreign leadership and employment practices imported from outside the Nordic context have placed pressure on established forms of cooperation and weakened collective representation. Together, these examples suggest that precarious labor is not only tolerated but increasingly accommodated within sectors historically embedded in the Nordic model.

At present, precarious employment remains marginal and often invisible, particularly in national accounts of health and safety. However, its growth reflects a tacit acceptance of new pockets of risk in working life. As precarious labor has become entrenched in many countries, it is unlikely to be reversed through incremental measures alone. Rather, adaptation through invisible and weakly protected forms of employment signals a shift toward living with precarization rather than actively confronting it. Our analysis resonates with what Bieder et al. (2024) describe as social or organizational hypocrisy: a situation in which continued advances in health, safety and wellbeing are secured for a privileged majority, while a marginalized group of workers experiences deteriorating conditions. Without deliberate efforts to address these structural cracks, the Nordic model risks gradual erosion rather than abrupt breakdown. This, in turn, calls for a renewed debate about whether existing frameworks are sufficient, or whether a different future must be envisaged for how health, safety, and wellbeing are governed in Nordic working life.

References

G. Alberti, I. Bessa, K. Hardy, V. Trappman, C. Umney, In, against and beyond precarity: working in insecure times. Work Employ. Soc. **32**(3), 447–457 (2018). https://doi.org/10.1177/0950017018762088

K. Alsos, S.M.N. Oppegaard, S.C. Trygstad, *Den norske arbeidslivsmodellen—status og utfordringer i en brytningstid [The Norwegian Model of Working Life–Status and Challenges in a Time of Change]* (Gyldendal, Oslo, 2024)

C. Bieder et al., *Industrial Safety in the Age of "Living With": Uncertainty, Complexity and Rising Expectations,* Number 2024–02 of the *Cahiers de la Sécurité Industrielle*, Foundation for an Industrial Safety Culture (Toulouse, France, 2024). https://doi.org/10.57071/420lwu

H. Bjørkhaug, A. Blekesaune, Gender and work in Norwegian family farm business. Sociol. Rural. **48**(2), 152–165 (2008). https://doi.org/10.1111/j.1467-9523.2008.00456.x

H. Bjørkhaug, J. Brobakk, Exceptional policies for exceptional situations? How COVID-19 revealed persisting precarity for seasonal migrant workers. Int. J. Sociol. Agric. Food **30**(1), 75–93 (2024). https://doi.org/10.48416/ijsaf.v30i1.571

B. Bungum, U. Forseth, E. Kvande (eds.), *Den norske modellen. Internasjonalisering som utfordring og vitalisering [The Norwegian Model. Internationalization as Challenge and Revitalization]* (Fagbokforlaget, Bergen, 2015).

H. Byrkjeflot, L. Mjøset, M. Mordhorst, K. Petersen (eds.), *The Making and Circulation of Nordic Models, Ideas and Images* (Routledge, London, 2022)

D.H. Constance, J.Y. Choi, M.K. Hendrickson, The southern model revisited: the intersection of race, ethnicity, immigration, and health and safety in poultry processing. Sustainability **15** (2023). https://doi.org/10.3390/su151813945

D. Côté, S. Durant, E. MacEachen, et al., A rapid scoping review of COVID-19 and vulnerable workers: Intersecting occupational and public health issues. Am. J. Ind. Med. **64**, 551–566 (2021). https://doi.org/10.1002/ajim.23256

J. Ernstsen, M. Varang, M. Pytte, *Utredning av ulykker, årsaker og tiltak i jordbruk, skogbruk og reindrift [Investigation of Accidents, Causes and Measures in Agriculture, Forestry and Reindeer Husbandry].* Main report ST-001655–1 (Safetec, Trondheim, 2025)

European Agency for Safety and Health at Work, *Precarious Work: Definitions, Workers Affected and OSH Consequences* (2024), https://oshwiki.osha.europa.eu/en/themes/precarious-work-definitions-workers-affected-and-osh-consequences

U. Forseth, Take-off, turbulence and turnaround. Scandinavian aviation. Equal. Divers. Equal. **42**(4), 480–493 (2022)

U. Forseth, Critical perspectives on work, health and wellbeing. Past, present and future. Chapt. 30, in *Handbook on the Sociology of Work*, ed. by C. Rees, Ö. Bozkust, S. Limoncelli, J. Preminger (Edward Elgar, Cheltenham, 2025)

U. Forseth, R. Rosness, Paradoxes of power: dialogue as a regulatory strategy in the Norwegian oil and gas industry. Saf. Sci. **139**, 105120 (2021). https://doi.org/10.1016/j.ssci.2020.105120

International Labour Organization (ILO) *Agriculture: A Hazardous Work* (2020), https://www.ilo.org/resource/agriculture-hazardous-work-0

International Labour Organization (ILO) *Construction* (n.d.), https://www.ilo.org/topics-and-sectors/industries-and-sectors/construction

A. Kalleberg, S.P. Vallas, Probing precarious work: theory, research and politics. Res. Sociol. Work. **31**, 1–30 (2018)

A.L. Kalleberg, *Precarious Lives. Job Insecurity and Well-being in Rich Democracies* (Polity Press, Cambridge, UK, 2018)

I. Koranyi, J. Jonsson, T. Rönnblad, L. Stockfelt, T. Bodin, (2018). Precarious employment and occupational accidents and injuries—a systematic review. Scand J. Work. Environ. Health **44**(4), 341–350 (2018). https://doi.org/10.5271/sjweh.3720

K. Lau, R. Aldridge, M. Norredam, G.F. Mkoma, M. Kugan, R. Chia-Yin Lin, L. Kiss, C. Zimmerman, S. Hargreaves, Workplace mortality risk and social determinants among migrant workers: a systematic review and meta-analysis. Lancet Public Health **9**(11), e935–e949 (2024). https://doi.org/10.1016/S2468-2667(24)00226-3

M. Nilsen, T. Kongsvik, P. Almklov, Splintered structures and workers without a workplace: How should safety science address the fragmentation of organizations? Saf. Sci. **148**, 105644 (2022). https://doi.org/10.1016/j.ssci.2021.105644

R. Rosness, U. Forseth, Boxing and dancing: tripartite collaboration as an integral part of a regulatory regime, in *Risk Governance of Offshore Oil and Gas Operations. Chapt. 12*, ed. by P.H. Lindøe, M. Baram, O. Renn, (Cambridge University Press, Cambridge, 2014)

J.F. Rye, M.H. Slettebak, H. Bjørkhaug, From family to domestic and global labour? A decade of proletarisation of labour in the Norwegian horticulture industry. Eur. Countryside **10**(4), 528–542 (2018)

Statistics Norway SSB, *Statistikk om arbeidsulykker* [Statistics on occupational accidents] (2024), https://www.ssb.no/helse/helseforhold-og-levevaner/statistikk/arbeidsulykker

H.N. Urrego-Parra, L.A. Rodriguez-Guerrero, R. Pastells-Peiró, The health of migrant agricultural workers in Europe: a scoping review. J. Immigr. Minor. Health **24**, 1580–1589 (2022). https://doi.org/10.1007/s10903-022-01330-y

K. Wasilkiewicz, E. Albrechtsen, S. Antonsen, Occupational safety in a globalized construction industry: a study on polish workers in Norway. Policy Pract. Health Safety **14**(2), 128–143 (2016)

K. Wasilkiewicz, S.S. Kilskar, A. Øren, R.K. Tinmannsvik, I. Kilanowska, Multicultural workplaces: a state-of-the-art study of the Norwegian construction industry, in *Safety and Reliability—Safe Societies in a Changing World*, ed. by S. Haugen et al., (Taylor & Francis Group, London, 2018)

I.M. Ydersbond, *The Organization of the Norwegian Aviation Sector in a European Context. TØI report 1869A* (Institute of Transport Economics Norway (TØI), Oslo, 2022)

GPSR Compliance

The European Union's (EU) General Product Safety Regulation (GPSR) is a set of rules that requires consumer products to be safe and our obligations to ensure this.

If you have any concerns about our products, you can contact us on ProductSafety@springernature.com

In case Publisher is established outside the EU, the EU authorized representative is:

Springer Nature Customer Service Center GmbH
Europaplatz 3
69115 Heidelberg, Germany

Batch number: 10799916

Printed by Printforce, the Netherlands